NanoScience and Technology

NanoScience and Technology

Series Editors:
P. Avouris B. Bhushan D. Bimberg K. von Klitzing H. Sakaki R. Wiesendanger

The series NanoScience and Technology is focused on the fascinating nano-world, mesoscopic physics, analysis with atomic resolution, nano and quantum-effect devices, nanomechanics and atomic-scale processes. All the basic aspects and technology-oriented developments in this emerging discipline are covered by comprehensive and timely books. The series constitutes a survey of the relevant special topics, which are presented by leading experts in the field. These books will appeal to researchers, engineers, and advanced students.

Please view available titles in *NanoScience and Technology* on series homepage
http://www.springer.com/series/3705/

Gianfranco Cerofolini

Nanoscale Devices

Fabrication, Functionalization,
and Accessibility
from the Macroscopic World

With 84 Figures

 Springer

Dr. Gianfranco Cerofolini
Università di Milano-Bicocca
Dipartimento Scienza dei Materiali
Via Roberto Cozzi 53, 20125 Milano, Italy
E-mail: gianfranco.cerofolini@mater.unimib.it

Series Editors:

Professor Dr. Phaedon Avouris
IBM Research Division
Nanometer Scale Science & Technology
Thomas J. Watson Research Center
P.O. Box 218
Yorktown Heights, NY 10598, USA

Professor Dr. Bharat Bhushan
Ohio State University
Nanotribology Laboratory
for Information Storage
and MEMS/NEMS (NLIM)
Suite 255, Ackerman Road 650
Columbus, Ohio 43210, USA

Professor Dr. Dieter Bimberg
TU Berlin, Fakutät Mathematik/
Naturwissenschaften
Institut für Festkörperphyisk
Hardenbergstr. 36
10623 Berlin, Germany

Professor Dr., Dres. h.c. Klaus
von Klitzing
Max-Planck-Institut
für Festkörperforschung
Heisenbergstr. 1
70569 Stuttgart, Germany

Professor Hiroyuki Sakaki
University of Tokyo
Institute of Industrial Science
4-6-1 Komaba, Meguro-ku
Tokyo 153-8505, Japan

Professor Dr. Roland Wiesendanger
Institut für Angewandte Physik
Universität Hamburg
Jungiusstr. 11
20355 Hamburg, Germany

NanoScience and Technology   ISSN 1434-4904
ISBN 978-3-540-92731-0   e-ISBN 978-3-540-92732-7
DOI 10.1007/978-3-540-92732-7
Springer Dordrecht Heidelberg London New York

Library of Congress Control Number: 2009929175

Printed on acid-free paper

Springer is part of Springer Science+Business Media (www.springer.com)

In memory of Mara Cerofolini (1952–2007)
scientist, teacher, and beloved sister

Preface

The second half of the twentieth century and the beginning of the twenty first have been characterized by the most impressive industrial revolution ever seen. In approximately 40 years, the complexity of integrated circuits (ICs) has increased by a factor of 10^9, with a corresponding reduction of the cost per bit by eight orders of magnitude.

Not only has this evolution allowed dramatic progress in all scientific fields (large computers, space probes, etc.), but also has fueled the economic development with the raise of new markets (personal computers, cellular phones, etc.) and even social revolutions (world wide web, global village, etc.).

In last years, however, the situation has significantly changed: the continuous scaling down of device size has eventually brought the IC major technique, photolithography, to its limits. Overcoming its original limits has been proved to be possible, but the price to pay for that has changed the playing rules – while at the beginning of the IC history the evolution was driven by technology, now it is driven by economy, the cost of a medium size production plant being in the range of a few billion dollars.

The predicted evolution is based on the assumption that future ICs will be based on the same basic constituent, the MOS-FET (metal–oxide–semiconductor field-effect transistor), suitably scaled to smaller and smaller sizes. The International Technology Roadmap for Semiconductors (the "Roadmap") has thus identified which progress is necessary to allow the increase of IC complexity to continue with the current exponential growth.

The number of researchers involved in the actions identified by the Roadmap is huge (about 10^5) and it is certainly possible that the limits it has identified will be beaten, but the cost of IC production plants is expected to increase by one order in a decade, putting it beyond the economic possibilities of most players.

In the light of these considerations, alternative solutions to the current top-down evolution of ICs have been considered. In particular, the fast development of nanotechnology (permitting the control of size effect, self-assembly, hydrophobic–hydrophilic properties, etc.) and the discovery of exploitable conformation, charge storage, and conduction properties of single molecules (π-conjugated moieties, rotaxanes, proteins, etc.) have attracted a large interest, suggesting the possibility of developing nanoelectronics in a different fashion – bottom-up.

In my opinion, in order to be viable that development requires the solution of the following problems:

- The design of new architectures based on electrically programmable molecules rather than on FETs
- The development of nonexpensive techniques for the synthesis, separation, and purification of such molecules
- The setup of an economically sustainable technology for the preparation of a distribution of active elements at a density (say, 10^{11} cm^{-2}) higher than the final one projected by the Roadmap
- The functionalization of such elements with the electrically programmable molecules
- The linkage of the functionalized elements to the external world to allow their actuation and sensing

An important point not considered in the above menu, and commented separately to emphasize it, is the temporal factor – *all the listed objectives must be achieved at a time appreciably closer than the end of Roadmap*. This need discards all alternatives not based on *conservative extension of the current technology*.

This book is written in that spirit: I intend to show that circuits formed by a planar arrangement of nanoscopic devices with density of the order of 10^{11} cm^{-2} can in principle be prepared via the combination of existing architectures, materials, and techniques. Whether or not that combination may indeed become a technology for nanoelectronics is essentially related to the effort spent in the solution of detailed problems that certainly will arise in such an attempt.

Although the temporal constraint does not apply to other devices (like sensors or electromechanical systems), their practical implementation would be greatly simplified if they too were based on some conservative extension of the current technology.

Although the topics considered in this book are highly technical, I have tried to maintain unity of style, privileging logical connections with respect to detailed information. Because of its formative, rather than informative character, this book (or at least its first part) might be used for courses on Nanoscience or Nanotechnology. A possible use of this book as a textbook is sketched in Fig. 0.1, where the backbone sketches the main body of the course, whereas the sides show possible additional topics.

The core of this book is based on two review articles: "Molecular electronics in silico," *Appl. Phys. A* **91**, 181 (2008), with E. Romano; and "Multispacer patterning: a technology for the nano era," to appear in *Handbook of Nanophysics*, edited by K. Sattler for Taylor & Francis (to be published), with E. Romano and P. Amato. In turn, they had origin from a series of lectures prepared for the 2007 and 2008 Summer Schools on Nanoelectronic Circuits at Tools held at the Ecole Polytechnique Fédérale de Lausanne (Swiss).

While acknowledging the contribution of my present coworkers [Paolo Amato (Numonyx) and Elisabetta Romano (Department of Materials Science, University of Milano–Bicocca)], I cannot, however, ignore the work carried out with my past collaborators [Clelia Galati, Sergio Reina, Lucio Renna, and Natalia Spinella

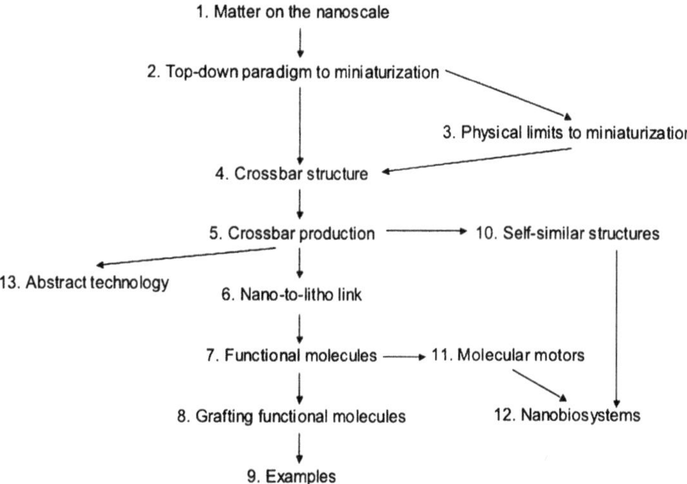

Fig. 0.1 Logical organization of the topics

(STMicroelectronics), whose activity was mainly addressed to the functionalization of the silicon surface, and Danilo Mascolo (STMicroelectronics) whose activity was addressed to the architecture of nanoelectronic circuits]. It is also a pleasure to thank a few colleagues (Eric Garfunkel, Rutgers University; Mark A. Reed, Yale University; and James M. Tour, Rice University) who read and commented the said article on molecular electronics. At last, I wish to express my thanks to Dario Narducci for supporting my position at the University of Milano–Bicocca where I completed this book, and to Hans J. Queisser for his warm suggestion of transforming my review paper on molecular electronics into a book.

Milano,   *Gianfranco Cerofolini*
June 2009

Contents

List of Acronyms

ADP	adenosine diphosphate
ATP	adenosine triphosphate
BET	Brunauer–Emmett–Teller
CMOS	complementary metal–oxide–semiconductor
CNT	carbon nanotube
CPU	central processing unit
CVD	chemical vapor deposition
DRAM	dynamic random-access memory
DUV	deep ultraviolet
EB	electron beam
EUV	extreme ultraviolet
FET	field-effect transistor
GSI	gigascale integration
HBT	horizontal beveling technique
HOMO	highest occupied molecular orbital
IC	integrated circuit
IL	imprint lithography
ITRS	International Technology Roadmap for Semiconductors
LOCOS	local oxidation of silicon
LSI	large-scale integration
LUMO	lowest unoccupied molecular orbital
mEMS	molecular electromechanical system
MEMS	microelectromechanical system
MIS	metal–insulator–semiconductor
MOS	metal–oxide–semiconductor
MSI	medium-scale integration
MWCNT	multiple-walled carbon nanotube
NEMS	nanoelectromechanical system
NLT	nonlithographic technique
nMOS	n-channel metal–oxide–semiconductor
NVM	nonvolatile memory
pMOS	p-channel metal–oxide–semiconductor

PSI	petascale integration
PVD	physical vapor deposition
QC	quantum confinement
RAM	random-access memory
RCA	Radio Corporation of America
SAM	self-assembled monolayer
SGS	Società Generale Semiconduttori
SNAP	superlattice nanowire pattern transfer
SOI	silicon on insulator
SPT	spacer patterning technique
S^nPT	multispacer patterning technique
S^nPT$_+$	additive multispacer patterning technique
S^nPT$_\times$	multiplicative multispacer patterning technique
SQC	smart quantum confinement
SSI	small-scale integration
SWCNT	single-walled carbon nanotube
TSI	terascale integration
VLSI	very large-scale integration
VPL	visible photoluminescence
WKB	Wentzel–Kramers–Brillouin

Part I
Basics

Chapter 1
Matter on the Nanoscale

Nanotechnology may be defined as the set of techniques[1] and methods for the exploitation of the properties of matter on the nanoscale.

Nanotechnology may also be considered as the practical realization of the dreams of Feynman and Drexler:

- In a legendary talk titled *There's plenty of room at the bottom: An invitation to enter a new world of physics*, delivered on 29 December 1959 at the Winter Meeting in the West of the American Physical Society [1],[2] Feynman proposed the use of atoms as information storage elements, thus allowing a planar bit density on the scale of 10^{15} cm^{-2} – the petascale integration (PSI).
- Subsequently, in a series of publications started in 1982 and culminating in a book on *Nanosystems – Molecular Machines, Manufacturing and Computation* [3], Drexler suggested that the optimal scale for the exploitation of microscopic objects at the macroscopic scale is the molecular or supermolecular one.[3]

Drexler's conclusion, although reducing the maximum density of a surface arrangement from the petascale to the terascale, does, however, clarify that the exploitation of the potentials offered by molecular or supramolecular arrangements requires the development of a technology for assembling, addressing and probing objects on the nanometer length scale – just nanotechnology.

---

[1] In this book the word "technique" is used with the meaning of "method of accomplishing a desired aim" (item 2b of Merriam–Webster Dictionary); "technology" will instead be used with the meaning of "combination of techniques aimed at a desired aim."

[2] A transcript of Feynman's talk is available online at http://www.zyvex.com/nanotech/feynman.html. In that site the talk is referred to as delivered to the "Annual Meeting," but Queisser informed the author that "it was the 'Winter Meeting in the West' of the American Physical Society in Pasadena. I was there, listened with enthusiasm to Feynman's physics and humor, even sat at the speaker's table; my boss Shockley arranged that for me. Shockley knew Feynman from his student days. The talk was very general, did not really address microelectronics as yet – but it provided a strong impetus for me" [2].

[3] The notion and even the name "nanotechnology" became, however, popular only after the appearance of *Engines of Creation. The Coming Era of Nanotechnology* [4], a visionary book listing the menu of what could reasonably be expected from a technology able to manipulate nanoscale objects.

1.1 Nanotechnology and the $(N + 1)$ Problem

The collective properties of matter can be distinguished in the light of their scaling behavior: on the *macroscopic scale* (the one described by classical physics) they remain substantially unchanged when the size of the considered system is progressively reduced; replicas of the system have negligible statistical dispersion; on the *mesoscopic scale* they change substantially when the size of the system is progressively reduced so that replicas of the systems have usually large statistical dispersion; whereas on the *microscopic scale* (the one described by quantum physics) matter cannot be shrunk and all replicas of the system are rigorously identical.

The upper limit of the microscopic scale is the molecular one, whose length scale is centered on the nanometer, so that mesoscopic systems have sizes typically in the interval 3–30 nm – that allows them, in relation to their size, to be referred to as nanoscopic. Not only do nanoscopic objects display properties (often potentially useful properties) generally different from those observed on the macroscopic scale, but also *on the nanoscale the properties of a system with $(N + 1)$ particles may be very different from those of a system with N particles.*[4]

The recognition that matter confined in geometries on a length scale of 3–30 nm manifests special behaviors has led to a systematic search for methods to permit their practical exploitation. The set of methods, processes, and materials allowing such exploitation forms the skeleton of nanotechnology.

Although eventually responsible for the birth of such a technology, the nanoscale property emphasized above, the so-called $(N + 1)$ *problem* [6], is responsible for the difficulties in the development of nanotechnology.

In perturbative terms the $(N + 1)$ problem may be restated saying that on the nanoscale the addition of another particle to a system of N particles cannot be considered as a small perturbation. Expressed in thermodynamic terms, it is equivalent to saying that the chemical potential of nanoscale systems is not a smooth function of N.

Clearly enough, the relevance of the $(N + 1)$ problem increases with the strength of the interaction of the $(N + 1)$th particle with the remaining N. The strength of this interaction is minimum when the particles are molecules interacting with each other by means of Van der Waals forces. Hence the *molecular road to nanotechnology*: exploiting the properties of packets of few molecules forming supramolecular complexes stabilized by Van der Waals interactions.

The molecular structure of matter was firmly established at the beginning of the twentieth century with the foundation of a theory of Brownian motion by Einstein and Smoluchowski and its experimental verification by Perrin. Molecules could however be individually sensed and manipulated only after Binnig and Rohrer's invention of the scanning tunnel microscope.

---

[4] Interviewed by the Editor for the first issue of *ACS Nano*, Rohrer defined nanoscale as the "bifurcation point where … materials have their properties and a cluster of 10 atoms does not yet have the same properties as 100 atoms" [5].

The availability of this tool has allowed the discovery that quantum properties of large potential interest (like interference or resonant tunnel) are typically manifested *inside* molecules, i.e., on the genuine nanometer length scale (0.3–3 nm).

Although one of the major aspects of current nanotechnology is just the functionalization of nanostructures with suitable molecules (e.g., for the lubrication of microelectromechanical systems, for drug delivery, for controlling hydrophobicity/hydrophilicity, etc.), the properties imparted by the molecules are qualitatively similar, although quantitatively superior, to those manifested on a macroscopic length scale – but really few devices exploiting the genuine intramolecular quantum phenomena are known.

1.2 Microelectronics is a Nanotechnology

An integrated circuit (IC) is a collection of batch-processed single devices [usually complementary metal–oxide–semiconductor (CMOS) field-effect transistors (FETs)] interconnected in such a way as to perform sophisticated operations, like computation. At the forefront of the technology, ICs have complexity on the gigascale integration; the gate length and width are of 45 nm, gate-insulator thickness is of 1–2 nm, and exploit, for some of their functions, genuine quantum phenomena (for instance, flash memories, with tunnel oxide thickness of 4–6 nm, are erased exploiting a typical quantum effect – Fowler–Nordheim emission). Thus, although usually referred to as *micro*electronics,[5] the IC technology is already a *nano*technology.[6]

Although quantum and ballistic phenomena certainly affect the behavior of metal oxide–semiconductor (MOS) FETs on a size scale of 10 nm, the analysis of test vehicles with that size has however shown that they can to a large extent continue to be regarded simply as large devices scaled down to that gate length. Which technology, if any, will be developed for the exploitation of such devices is still obscure; rather, there is a general agreement in the field that some of the fundamental difficulties in developing devices scaled to the 10-nm length scale will result in a failure of the technology within one decade. This consideration suggests the idea of developing a radically new technology exploiting (rather than trying to cancel) the properties manifested on the nanometer length scale.

---

[5] Actually, prefix "micro" in the present context may be deceptive: affixed to "meter," it denotes a *macro*scopic length scale; affixed to "electronics," it denotes a *nano*technology; while affixed to "scopic," it is almost synonymous with "quantum." To avoid the confusion deriving from its use, from now on the objects forming the ICs will be referred to mentioning the basic technology involved in their production – lithography.

[6] According to the United States Patent Office, for an invention to qualify as pertaining to nanotechnology, at least one of its dimensions should be between 1 and 100 nm in size, and the tiny size of the device must be essential to its function [7].

Macroscopic devices based on nanoscopic elements have already been developed (think, for instance, of quantum dots for single electron transistors). The $(N + 1)$ problem, however, makes their reproducibility extremely difficult. This fact makes difficult the preparation of dense arrays of such devices with uniform characteristics.

However, the size of the cluster, at which one property, say the energy gap, is half way between the bulk and individual atom, will differ from one material to the next, and depend strongly upon electronic bonding or coupling. Whereas strong coupling will result in nonsmooth dependence on N, weak coupling, like that occurring in noble gases and to a lesser extent in molecular solids, will result in smaller differences between the individual molecule and the cluster. A radically new approach to nanoelectronics is thus based on the exploitation of conduction or charge-storage properties of single molecules or of packets of few of them.

1.3 From Microlectronics to Molecular Electronics

The birth of molecular electronics can be dated with the appearance of Aviram and Ratner's seminal paper [8]; that work opened the road to the search for and determination of molecular properties (including strongly nonlinear conductance, bistability, etc.) that have analogues in electronic devices. Attention was correspondingly focused on the preparation of simple test vehicles for checking whether a molecule, or an assembly of a few molecules, manifests properties of interest for electronic applications. In turn, this focus has led to experiments (involving scanning tunneling microscopy, break junctions, etc.) for the determination of such properties. This state of affairs is well described by the following words by Reed [9]:

> We are still [2001] at the materials stage in this field [molecular electronics], working to understand fundamentally what is going on in the materials by measuring their transport characteristics. Electronics has to go one step further, to a point where we can actually interconnect all of the devices. How we are actually going to progress technologically to the point where we can connect molecules is still not clear... Learning how we can connect all of these structures at the molecular stage is the next stage.

Transforming discrete molecular *devices* into practically exploitable *circuits* (where billions of molecular devices work in concert and each of them is electrically accessible) indeed requires the development of a totally new technology. As emphasized by Tour [10, p. 251], a silicon–organic hybrid architecture seems an obvious solution:

> with the fundamental limits in solid-state structures being approached, molecules will become the devices of the future ultradense computational systems. Not only will they be smaller and less expensive than their solid-state counterparts, but in some instances they will show superior behavior which is hitherto unattainable from solid-state devices. The key, however, is that molecular systems will likely not supplant solid-state systems, but that they will be complements to traditional electronic devices.

The path to molecular electronics, however, is not expected to be so easy. In a critical analysis, Keyes [11, p. 20] objected that as molecular electronics requires the development of some novel architecture, "history does not encourage optimism." Also, the technology requires the solution of a number of problems [11, p. 23]:

> Strains from attachment to a substrate will have an effect; inhomogeneities in a substrate surface may also modify properties. Gradients of temperature will create differences between different regions. Also, one must ask what is necessary to ensure that all of a large number of molecules actually do have the same chemical composition. Quantitative questions will arise. What standards of reagent purity will be required in order to obtain satisfactory yield in an assembly of 10^6, 10^8, or 10^{10} molecules? What sort of clean environment will be wanted to control the contamination that plagues electronic manufacturing?

None of the above questions has been answered yet.

Assuming optimistically that the concerted research effort in this area will succeed in giving positive answers, in this book the attention will be concentrated on (although not limited to) the molecular property which has attracted major interest for its possible implementation in hybrid circuits – the dramatic change of conductance resulting from the excitation of an ionized donor–acceptor pair [12] or from the ionization of a suitable molecular center [13]. If the excited states are metastable, the molecular bistability can be exploited for the preparation of memories.

Chapter 2
Top-Down Paradigm to Miniaturization

The current development of electronics is ultimately a consequence of the invention of the MOS-FET. The idea upon which the FET is based, of controlling the conductivity of a region by varying its surface potential, precedes by many years its practical realization and industrial implementation. In fact, the invention of the metal–insulator–semiconductor (MIS) FET goes as far back as to 1926 for the depletion-mode FET [14] and to 1935 for the enhancement-mode FET [15]. Quite ironically, in an attempt to produce MIS-FETs on semiconducting germanium, Bardeen, Brattain, and Shockley discovered the bipolar transistor in 1947 (an insider view of the history of the transistor discovery[1] is given in [16]). It was only after the change of substrate (from germanium to silicon) that the first FET could actually be prepared (by Khang and Atalla in 1960 [17]) by changing the insulator to SiO_2 and the semiconductor to silicon – and the MOS-FET opened the main avenue to the IC era.[2]

Most ICs are indeed constituted by MOS-FETs, connected in a way that allows assigned electrical functions to be performed – the more sophisticated the functions, the more complex the circuit. The early IC market was however slow to develop and its exponential growth was triggered by Fairchild's decision to sell IC chips at a price lower than the sum of the prices of the individual components necessary to make an equivalent circuit [20].

The cost reduction was possible thanks to the possibility of scaling down, in a relatively easy way, the size of the elementary devices (bipolar transistors first and FETs later) forming the circuit. Figure 2.1 compares an MOS transistor of the mid-1980s with one produced 20 years later: although the zone outside the crystalline silicon are remarkably different, channel, source, and drain may be seen as scaled copies of one another.

---

[1] Whether the preparation of the transistor should be considered an invention (in view of its engineering content) or a discovery (in view of the amount of serendipity involved in the preparation) is a matter of debate.

[2] The short introduction above is not intended to present a historical view of the development of microelectronics. A friendly overview of the golden age of microelectronics is given in [18]; a more specialized history devoted to the microprocessor is given in [19].

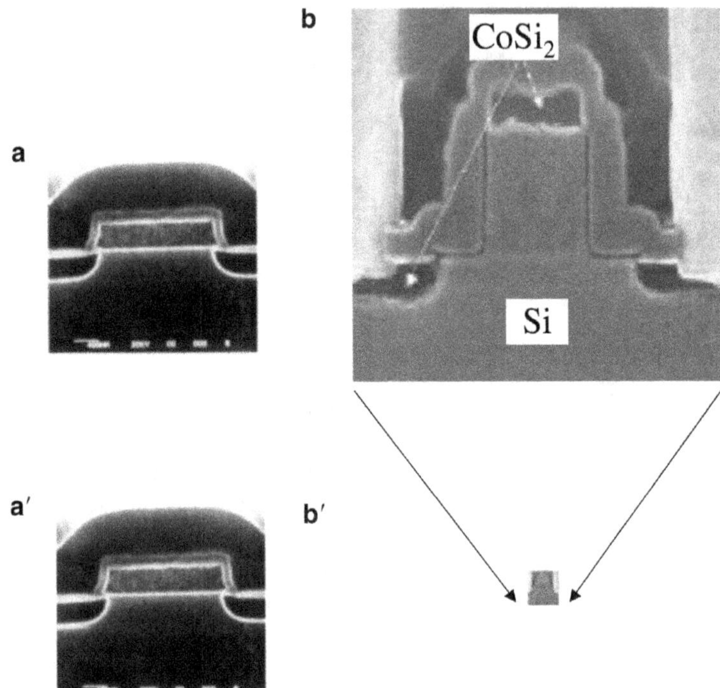

Fig. 2.1 Comparison of the cross sections of two MOS-FETs prepared with the technology of the mid-1980s (*left*) or with current technology (2005) at the forefront (*right*). In the upper part, the length scales have been arranged so as to have the transistors (**a**) and (**b**) with the same channel length, whereas in the lower part the same length scale was used for both transistors (**a′**) and (**b′**)

2.1 The Path Toward Size Reduction

The development of IC technology in the last 40 years is described by a set of statements collectively known as Moore's laws; its evolution is usually predicted assuming that those statements will continue to hold true in the near future too.

Although predicting the future is a notoriously risky activity, there is indeed a widespread consensus that the concerted activities of research and development outlined by the International Technology Roadmap for Semiconductors (the "Roadmap") guarantees that the evolution of ICs will continue in the next 10 years as described by Moore's laws.

Moore's first law is a statement describing the increase with time of circuit complexity. It has substantially remained true over approximately 40 years [21], except for a modest correction to the original doubling time [22] and its specialization for different ICs:

Moore's First Law.   *The number of transistors on integrated circuits doubles every 18 months for memories, or every 24 months for microprocessors.*

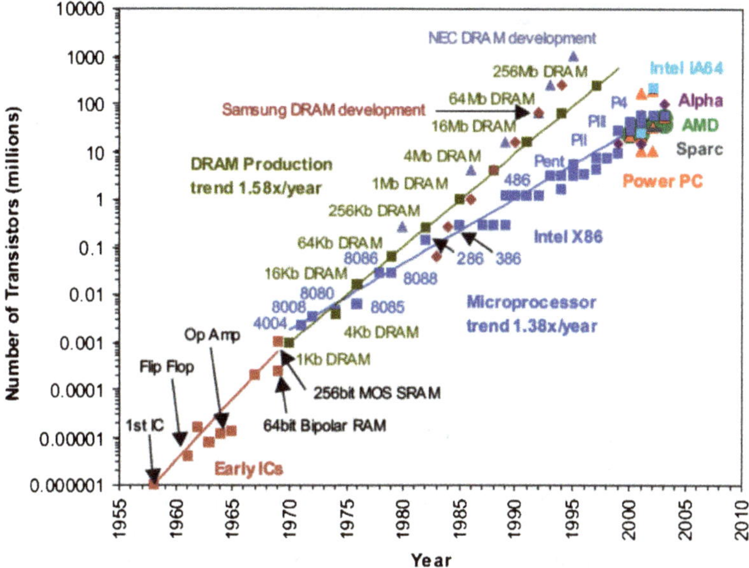

Fig. 2.2 The Moore's first law: time variation of the number of transistors in ICs (reproduced with permission from http://www.icknowledge.com)

Figure 2.2 supports this statement.

The process of shrinking devices and increasing IC complexity has never been trivial and several seemingly insurmountable bottlenecks have been overcome. This impressive revolution has been possible thanks to the setting-up of the IC technology – a technology essentially based on the combination and repetition of a few basic techniques, in turn requiring the use of relatively few materials (primarily Si, SiO_2, S_3N_4, Al, $TiSi_2$, TiN, and W, in addition to sacrificial materials like photosensitive polymers and gas or liquid etchants).

Combining the shrinking of FETs with the development of IC technology, the evolution of microelectronics has been dominated by the MOS-FET paradigm, where "MOS-FET" is used, here and in the following, with the restrictive meaning of "metal–oxide–*silicon* FET":

MOS-FET Paradigm. *Families of denser and denser devices are built scaling down silicon-based FETs.*

It has become a common usage to refer to the process of scaling down large circuits to produce similar, still working, smaller devices as "top-down" route.

The current integration has been obtained through a complex evolution in which families of technologies are classified in relation to their ability to achieve a certain class of integration. An inspection of the major changes associated with the evolution of microelectronics shows that, rather than the use of materials or devices with better intrinsic performances, the passage from one generation to another was obtained with a continuous improvement of the photolithography and with the

Table 2.1 Complexity of semiconductor integrated circuits

Year	Class	Bit	Size		
			Transistor (μm)	Chip (mm^2)	Wafer (mm)
1960–1968	SSI	2–128	> 10	1–15	25
1965–1975	MSI	64–4 K	3–10	10–25	50
1972–1983	LSI	2 K–128 K	1.5–4	15–50	50–100
1980–1988	VLSI	64 K–4 M	0.75–2	25–75	100–125
1985–1993	ULSI	2 M–128 M	0.5–1	50–200	125–200
1990–1999	GSI	64 M–4 G	< 0.5	100–400	$\geq$ 200

SSI, *MSI*, *LSI*, *VLSI*, *ULSI*, and *GSI* mean small-, medium-, large-, very large-, ultra-large-, and gigascale integration, respectively

invention of techniques [like the silicon-gate technology,[3] local oxidation of silicon (LOCOS),[4] and spacer patterning technology (SPT)[5]] that allow the self-alignment, rather than the lithographic alignment, of a layer with respect to an underlying pattern.[6] This virtuous combination has allowed the mass production of FETs with feature size (gate length) of 45 nm and complexity on the gigascale integration (GSI). Table 2.1, taken almost *verbatim* from a work published in 1984 [30], summarizes the first 40 years of IC evolution. For instance, the small-scale integration (SSI, characterized by the number of bits per device in the interval 2–128) was obtained by means of bipolar technology; the medium-scale integration (MSI) was possible either by bipolar technology or by the aluminum gate, p-channel, MOS technology; the large-scale integration (LSI) was characterized by the silicon gate, local oxidation of silicon (LOCOS), n-channel, MOS technology (nMOS). There is a general consensus that the MOS technology will continue to be the leading technology in the near future too.

The increase of complexity has been possible thanks not only to a more sophisticated design but also to a progressive decrease of the minimum producible feature size. The achievement of the GSI has required the development of suitable lithographic techniques for the definition of shorter and shorter FETs.

Preparing denser and denser ICs is however a difficult and expensive job. The industrial system has succeeded in that, but the cost of fabrication facilities ("fab") has increased exponentially too:

Moore's Second Law. *The average cost of fabs for manufacturing ICs increases by a factor of 2 every 3 years.*

---

[3] The technique for the simultaneous definition of gate and source-and-drain was invented in Fairchild by Faggin and coworkers [23, 24]. However, it became a widespread technology only after the industrial scale up in INTEL [25].

[4] The technique of local oxidation of silicon was invented in Philips [26, 27] and almost simultaneously in SGS (now STMicroelectronics) [28].

[5] As a technique for the self-alignment of contacts with respect to source and drain, the SPT was invented in INTEL [29].

[6] A layer is said to be *self-aligned* with respect to predefined patterns when its pattern is obtained without any masking procedure.

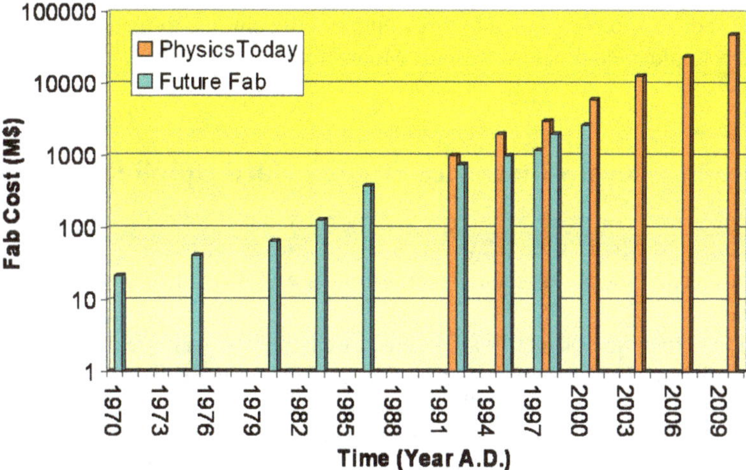

Fig. 2.3 The Moore's second law: time variation of the cost of fabrication facilities for IC manufacturing. Upper data are projections from [31] and lower data are a survey from [32]

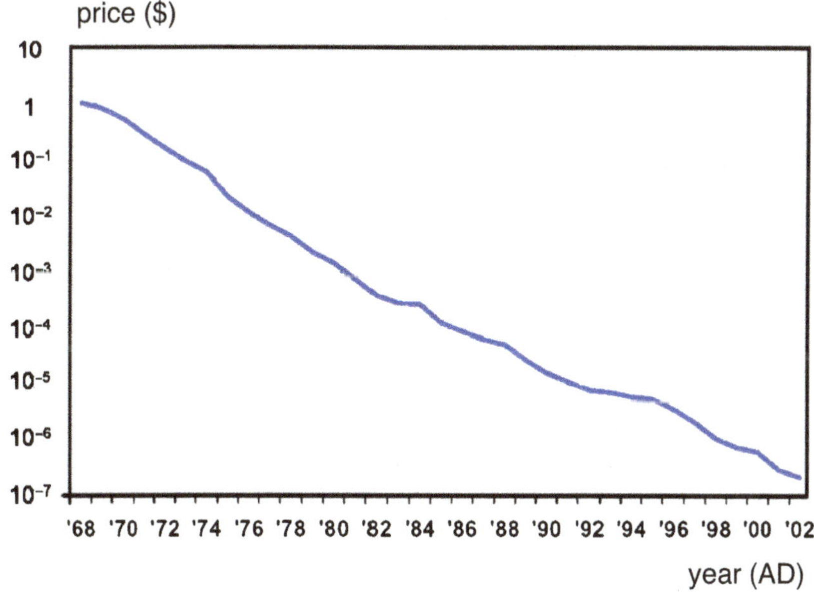

Fig. 2.4 Time evolution of the bit price (taken from [33])

Figure 2.3 describes this trend pictorially.

The increase in productivity resulting from the combination of the Moore's first and second laws has allowed the microelectronic industry to reduce the average price of MOS-FETs by one order of magnitude in an average period of 5 years (see Fig. 2.4). Actually, as already mentioned, the engine of the information revolution has just seen this decrease, and that revolution will continue to be fueled as long as the IC technology continues to respect Moore's laws.

2.2 Going Down with Device Size is a Hard Uphill Path

Of course, the Moore's first law cannot hold true indefinitely – sooner or later physical, technological, or economic constraints will limit the effort toward higher integration.

The Roadmap is essentially a Gantt chart[7] listing the major technical improvements which are expected to be necessary to maintain the validity of the Moore's first law in the next few years [34]. The Roadmap has predicted the introduction of the 45-nm-node MOS technology into mass production by 2010. This will require the technological capability of producing and integrating FETs with gate length shorter than 25 nm.

There are only feasibility indications beyond the limits of the Roadmap, and no corresponding technology has so far been hypothesized for practical productions. Even limiting the attention to the goals of the Roadmap, it is not clear whether or not they can actually be achieved in an economically sustainable way.

2.2.1 The Physical Limit

The mainstream of IC research in the past decade has been addressed to redesign the FET core (high-K dielectrics as gate insulators; K denotes the dielectric constant) and the FET interconnects (low-K dielectrics for intermetal insulators and copper for long-range metal conductors). In the absence of superconductors at room temperature, the physical limits for interconnects have almost been achieved.

The *physical limit* is related to the nonscalability of the MOS-FET below a critical size L_{phys}. This limit denotes here not only the failure of the transistor model upon which the MOS-FET paradigm is based, but also the fact that the main MOS-FET electrical characteristics (like the infinite channel-gate impedance, the negligible current in OFF condition, and the saturation current in ON condition – features which have allowed the integration of more and more functions in a unique circuit like a microprocessor) are lost below L_{phys}. Of course, these failures do not necessarily imply that it is not possible to construct electronic circuits with such elements (this use seems conceptually possible down to a size scale where nonlinear current–voltage characteristics are preserved), but rather that the circuit will

---

[7] A Gantt chart is a time schedule of a project, illustrating the start and finish dates of the terminal elements and summary elements. Terminal elements and summary elements comprise the work breakdown structure of the project.

progressively lose its digital behavior to assume an analogic character – that notoriously makes difficult large-scale integration. Designing complex circuits with linear rather than digital devices is indeed a major difficulty, which has eventually led to the almost complete occupation of the IC market with digital circuits.[8] On a length scale down to 20 nm, the conventional MOS-FET architecture is expected to fail as the control over short channel effect as well as its current-driving capability is lost. To make up for these difficulties new devices inspired by MOS-FETs, but which might theoretically be useful for even harder geometries (in the range of 10 nm), have however been proposed [37]. The consideration of these new devices leaves open the problem of which new factors will ultimately limit the transistor performances.

2.2.2 The Technological Limit

The *lithographic limit* is related to the impossibility of batch defining, using proximity masks, feature sizes below a critical one L_{litho}.

The top-down route of the IC technology had been relatively easy to run until its basic step (photolithography) met its first physical limits (minimum feature size around the light wavelength). Further improvements have been possible only thanks to masterly combinations of reticle enhancement techniques (optical proximity correction, phase shifting masks, and off-axis illumination) which have allowed (in combination with techniques like the resist-ashing and oxide-trimming process steps) the definition of geometries with size lower than the wavelength by a factor of 2–3. Figure 2.5 shows how the lithography has succeeded in the production of feature size appreciably shorter than the wavelength.

In more detail, generations of lithographic techniques are usually classified according to the medium required for the definition of the wanted features on photo- or electrosensitive materials (resists): standard photolithography (436 nm, Hg g-line; refractive optics), deep ultraviolet (DUV) photolithography (193 nm, ArF excimer laser; refractive optics), and immersion DUV photolithography (refractive optics). Figure 2.5 sketches the time evolution of the lithographic techniques.

---

[8] The comparison of the shrinking rate of MOS devices with that of new functions added to the IC shows that the first rate is much higher than the second one. In fact, the increase of design complexity with the number of constituting devices has led to progressively less efficient design solutions, favoring the design time rather than the bit area. In the diagram complexity vs. time the effect of this choice is reflected in the different growth rates for repetitive and simple structures (like memories) and for complex circuits (like microprocessors). The *design gap*, between the number of technologically integrable devices and the number of those practically integrated, is remarkable (compare in Fig. 2.2 the evolution of memories and microprocessors) and a major progress in the methodology of automatic design could increase by one order of magnitude the integration density with no new technological efforts [35, 36].

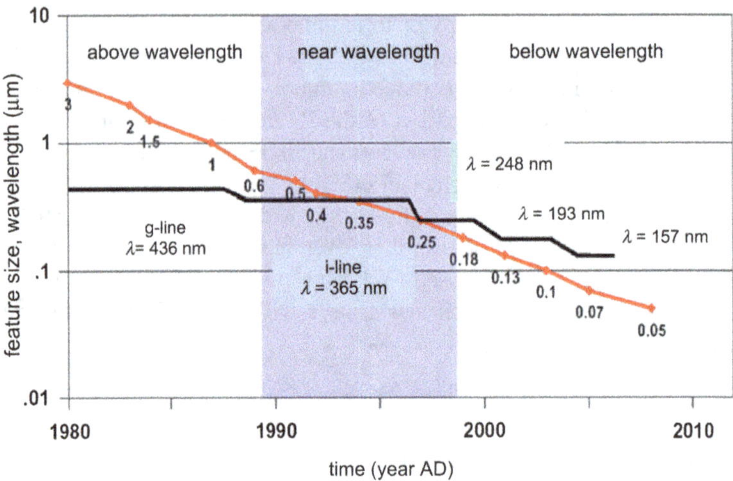

Fig. 2.5 Temporal evolution of the technologically attainable feature size and the wavelength allowing its definition

Most likely such improvements will continue in the near future too, but only using new exposure sources (extreme ultraviolet, synchrotron radiation, or electron beam) and extending the reticle enhancement techniques to the new wavelength region.

2.2.3 The Economic Limit

Techniques for achieving the 10-nm length scale are known – extreme ultraviolet (EUV) photolithography (13.5 nm, plasma-light source; reflective optics) and electron beam (EB) lithography (electron wavelength controlled by the energy, typically in the interval 10^{-3}–10^{-2} nm). Their industrial application, however, requires huge investments because of either the required investment per machine,

$$\text{DUV} \ll \text{immersion DUV} \ll \text{EUV},$$

or the throughput,

$$\text{EB} \ll \text{DUV}.$$

The dramatic explosion of the cost required for the reduction of the feature size is sketched in Fig. 2.6. The trend shown therein has thrown doubts on the possibility of continuing the current increase of density.

Combining these costs with the ones produced by the increase of process complexity (see Fig. 2.7) generates the *economic limit* L_{eco}, i.e., the minimum device size allowing a return of the investment.

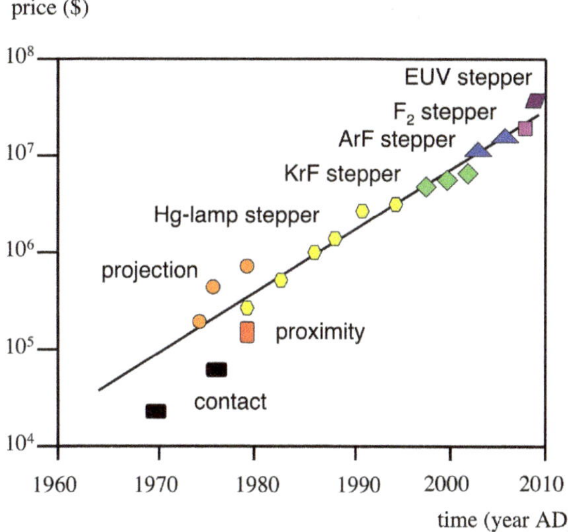

Fig. 2.6 Exponential increase with time of the cost of lithographic systems

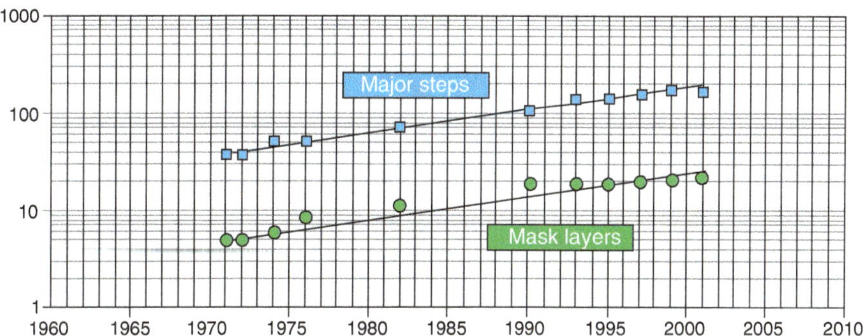

Fig. 2.7 Exponential increase with time of the complexity of the process required for the production of logics (reproduced with permission from http://www.icknowledge.com)

2.3 Going Beneath the Limiting Size

Let L^* denote the maximum among L_{phys}, L_{litho}, and L_{eco}. At present, it is difficult to make a sensible prediction about the value of L^*. An upper limit is suggested by the Roadmap: $L^* \lesssim 30$ nm.

Assuming a physical limit of 10 nm, in view of the above considerations, L^* is in the interval 10–30 nm. The closer L^* is to the upper limit of this interval, the sooner the IC industry will become a mature industry and the wider is the room for the development of new (at present hypothetical) ICs where the data handling elements are single molecules.

In molecular electronics the preparation of single devices forming the circuit is left to the chemist, and involves steps with which he or she is familiar: molecular design and modeling, synthesis, separation, purification, etc. The IC preparation requires the arrangement and interconnection of these molecules to achieve an assigned electrical function – that requires the development of a new *bottom-up* technology.

Chapter 3
Physical Limits to Miniaturization

Understanding the ultimate scale below which a device no longer behaves as extrapolated from its behavior on a larger scale is not trivial. A special case, of topical importance, is understanding which are the physical limits below which electronic computation is impossible.

Most ICs are indeed produced for handling data. They are usually classified into two families, *logics* and *memories*, according to the way they operate in combinatorial or sequential way. From the architectural point of view, the logic gates are arranged and linked in a way controlled by the specific logic functions they implement; on the contrary, memories are organized in a periodic arrangement of their constituting cells.

In view of their greater architectural simplicity, the limits of computation are better understood considering memories. Though restrictive, not only is the forthcoming analysis addressed to the largest class of electron devices, but also to the one that is expected to meet such limits sooner because of the fastest growth rate in the increase of complexity.

Following the analysis of [38], the ultimate limits of computation will be determined in the hypotheses that computation is a physical process occurring in a medium immersed in a thermal reservoir at an assigned temperature and thus obeying the underlying physical laws. The computational figure of merit will be calculated considering the need for transforming the microscopic computation outcome into a macroscopic event.

3.1 A Case Study: The Limits of Computation

Even though the race toward larger and larger complexity of ICs silicon has met some "physical limits," their impact has never been so strong as to slow down the exponential growth of microelectronics. The current rate of integration is characterized by an increase of bit density by almost one order of magnitude each 6 years; this rate would bring electronics into the TSI era in approximately 20 years, were it not for the fact that this density is incompatible with the physics of the basic constituents of ICs (the MOS-FETs) and with the ways they are mutually interconnected and connected to the external world [39].

Speculating on which are the ultimate density limits of circuit integration is an activity involving fundamental principles of quantum and statistical mechanics (classical and quantum computation; macroscopic detection of microscopic states; mutual relationships between entropy, information, and dissipation; etc.) and has been the matter of extended investigations [40–52]. The conclusions of such a study may also impact the perspectives of silicon technology because they address the problems which are expected to be met with the progressive shrinking of device size (see [53–56] for a discussion of the principles that can be exploited, within the frame of the current technological paradigm, for new devices).

In the last few years, there have been numerous efforts to remove the constraint of using current technology or devices: recent experiments on condensed-matter-based systems (with characteristic size on the super-nanometer length scale) for quantum-information processing have shown that they also can carry information, for instance, in macroscopic quantum states (superconducting charge or flux q-bits) or in charge or spin states of quantum dot. It is certainly possible that overcoming the classical Turing machine by new paradigms [not only with intrinsically higher parallelism (quantum computers), but also with learning (neural networks) or fault tolerance (bioinspired systems)] will render vain the current effort toward larger and larger integration. This chapter is however based on the stipulation of the persistence of the current computational paradigm.

Within this *conservative* approach (for which computation is essentially a dissipative irreversible process), the analysis will be:

- *Application-driven*, having in mind devices operating at room temperature[1]
- *Concrete*, the attention being concentrated on a device which, though highly ideal, has features that resemble the ones of a real device – the nonvolatile memory (NVM) cell
- *Technology-oriented*, being limited to memory cells arranged in planar configurations

Of course, removing one or the other of the above constraints (allowing, for instance, operations close to 0 K, three-dimensional arrangements, etc.) would lead to different conclusions.

3.2 The Basic Computational Unit

The figure of merit of a planar memory is the flop rate per unit area, $\dot{F}$, which should be as high as possible:

$$\dot{F} = \max.$$

---

[1] Speaking of fundamental limits, the condition of operating at room temperature may seem a ridiculous constraint. However, it is an essential condition when dealing with large-volume practical applications. This constraint is so important that, according to Stoneham, even quantum computation "should work above cryogenic temperatures, and perhaps at room temperature" [57].

Since $\dot{F}$ may be written as the product of the cell switching frequency ν times the cell density ρ,

$$\dot{F} = \nu\rho,$$

the problem of maximizing $\dot{F}$ is reduced to the problem of finding the combination of ν and ρ satisfying the following condition:

$$\nu\rho = \max.$$

Needless to say, computation must satisfy the underlying physical laws that are specialized according to the size of the cell, its material structure and environment, and how it is linked to the rest of the world. If the attention is focused on the dynamics, the limits are presumably dictated by the uncertainty principle and the ballistic motion of the information carrier; specifying the materials of which the cell is formed will introduce constraints related to the material properties; if the cell is embedded in a thermal reservoir at temperature T, decoherence phenomena and statistical properties ultimately related to the Second Law of thermodynamics must be taken into account; and considering the phenomena (semiconductivity, superconductivity, etc.) exploited for computation will result in limits given by the macroscopic laws governing those phenomena. At the current level of development, the effort toward larger and larger integration has been mainly limited by the last kind of constraints. However, they are related to specific choices of materials and technology and will henceforth be ignored.

As far as computation is indeed an irreversible process, computers can be thought of as engines for transforming free energy into mathematical work and waste heat. Even assuming that the thermodynamics of computing is adequately known (see however Sect. 3.4), it however does not provide any information on the computing speed [45, 47]. Predicting how fast a computer is in performing the calculation (a problem totally ignored by thermodynamics) requires a minimal dynamical model of the computing act.

The forthcoming analysis is based on the fundamental assumption that the cell is constituted by a single object (the "information carrier") of mass m; the dynamics of this object is described by the Schrödinger equation in a symmetric potential with two equivalent wells separated by a barrier of height ΔE_{bit}. Moreover, to simplify the calculations all potentials are assumed to be sufficiently smooth to allow the use of classical mechanics to describe the ballistic motion of the carrier (Ehrenfest theorem), provided that the quantum expectation values are substituted for the corresponding classical variables (in this situation, tunneling from one minimum to the other is disallowed).

The information carrier is assumed to be localized around one or the other of the potential minima. The state vectors describing such states, $|0\rangle$ and $|1\rangle$, may be used to code two *perfectly distinguishable* states only if they are orthogonal. This implies that the supports of $\langle q|0\rangle$ and $\langle q|1\rangle$ (with q a position coordinate) are not overlapped.[2] The extension of the supports of $|0\rangle$ and $|1\rangle$ is a measure of the position uncertainty.

Since the potential has two equivalent minima, if the particle is in an energy eigenstate it is spread over both minima; since the states $|0\rangle$ and $|1\rangle$ are localized each around a potential minimum, they are necessarily combinations of energy eigenstates with dispersion ΔE.

The basic idea is that the system may be prepared in one or other of two[3] eigenstates $|0\rangle$ or $|1\rangle$, where the carrier is spread around one or the other of the potential minima. Thus $|0\rangle$ and $|1\rangle$ are eigenstates neither of position nor of energy. They are nonetheless almost orthogonal and may be thought to be a basis of "logic states."[4] Since they are not energy eigenstates, they necessarily vary with time; however, the potential can be chosen in such a way as to allow their persistence in the absence of external potential for a macroscopically long time. In the absence of any other phenomenon, the information is unavoidably lost via spreading of the wave function over the entire cell. Denoting with t_{spread} such a time and with ℓ_{cell} the cell size, this time is approximately given by

$$t_{\mathrm{spread}} \approx 2m\ell_{\mathrm{cell}}^2/\hbar, \tag{3.1}$$

where $\hbar$ is the reduced Planck constant. Under certain conditions, however, spontaneous decoherence phenomena reduce the quantum logic states $|0\rangle$ and $|1\rangle$ to classical states, 0 or 1, which in the absence of interactions with the environment do not vary with time. The cell is assumed to undergo two possible operations:

Programming, the confirmation of the original state, or the switching to the other one

sensing, the detection of the actual state of the particle

In this way the memory cell resembles the cell of nonvolatile memories.[5]

---

[2] Of course, unless infinite potentials are considered, some degree of overlapping is even conceptually unavoidable; the inner product,

$$\langle 0|1\rangle = \int_{\mathrm{space}} \mathrm{d}q\, \langle 0|q\rangle\langle q|1\rangle,$$

may thus be taken as a measure of the intrinsic inaccuracy of any physical computation.

[3] Actually even this assumption is a strong restriction. The use of three or more logic levels is by itself a way of increasing the *bit* density more than *cell* density.

[4] The necessarily unavoidable partial superposition of the supports of $\langle 0|q\rangle$ and $\langle 1|q\rangle$ might in principle be exploited for the practical realization of a three-level logic with false, true, and indeterminate values.

[5] Of course, the similarity must not be taken in a strict sense. In CMOS nonvolatile memories, states 0 and 1 are classical macroscopic states; programming is achieved either via the injection of 10^2–10^4 electrons into a floating gate from the underlying channel ($1 \rightarrow 0$: WRITE) or via their field-assisted tunneling to the source ($0 \rightarrow 1$: ERASE). Moreover, whereas erasing is a relatively efficient process involving the dissipation of approximately 1 eV per electron, the efficiency of writing depends on memory architecture: for NOR memories writing is an inefficient process (yield

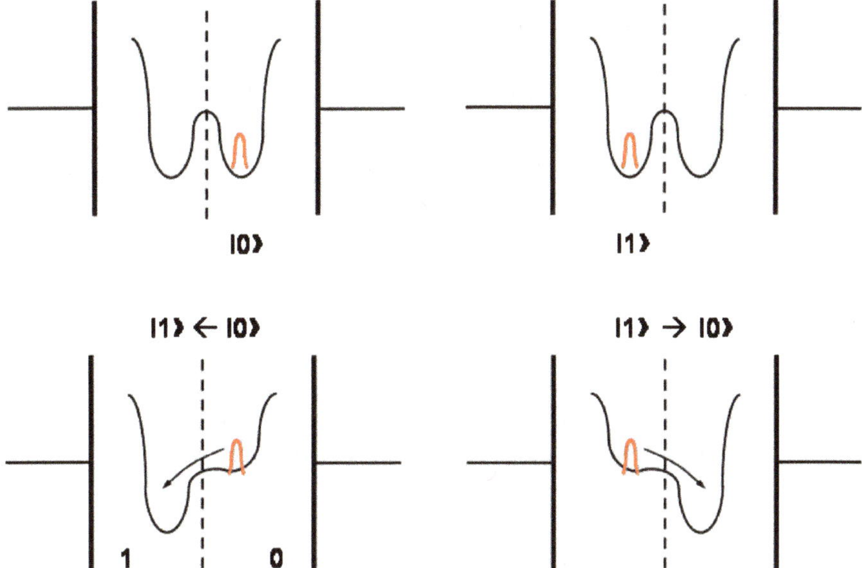

Fig. 3.1 Programming the cell: from $|0\rangle$ to $|1\rangle$ (*left*) or from $|1\rangle$ to $|0\rangle$ (*right*)

Even though this model is highly ideal and real computation requires certainly more complicate arrangements, it is nonetheless often considered and is at the basis of several fundamental investigations [40, 48]. In this chapter, without anyway specifying the structure of the ultimate memory cell, it will be assumed to be characterized by the same operation modes that characterize nonvolatile memories: PROGRAM and READ.

Programming is achieved decreasing the potential energy of one well, in such a way as either to stabilize further the carrier therein if already there, or to produce the drift of the carrier from the other well. The application of such a potential is assumed to last a time sufficiently long to allow the dissipation of heat of the original particle energy. In this way any programming act implies the dissipation of an energy of about ΔE_{bit}. Figure 3.1 sketches the situation.

Sensing is essentially a measurement process where the state of the carrier is detected. The theoretical description of sensing is a difficult point because it is intrinsically associated with the measurement of a microscopic state – a foundational problem that has so far had partial and unsatisfactory solutions only (see, for instance, the collection of papers in [58]). However, under certain circumstances (discussed in Sect. 3.3.3) the microscopic state may be regarded as classical, that allows its determination without running into the difficulties of the measurement problem.

---

of the order of 0.1%) involving "hot" electrons (with kinetic energy of 3–4 eV), whereas for NAND memories writing is a relatively efficient process occurring via field emission from the channel.

3.3 Programming

3.3.1 Limits Imposed by the Uncertainty Principle

The uncertainty principle sets limits to both the minimum switching time and maximum density.

Switching Time

To avoid a spontaneous loss of information, the energy dispersion ΔE of a state cannot be larger than the energy ΔE_{bit} required for switching from the corresponding logical state to another, $\Delta E < \Delta E_{\text{bit}}$. The minimum time required by any system to switch to a distinguishable (thus orthogonal) state is given by Heisenberg relation [59]:

$$\Delta t \geq \hbar/\Delta E.$$

The minimum time $\Delta t_{\min}$ is achieved when $\Delta E = \Delta E_{\text{bit}}$, so that $\Delta t \geq t_{\min}$, with

$$\Delta t_{\min} = \hbar/\Delta E_{\text{bit}}.$$

A complete switching cycle is obtained when two transitions, $|0\rangle \to |1\rangle \to |0\rangle$ or $|1\rangle \to |0\rangle \to |1\rangle$, have occurred. Thus, the switching frequency of a single memory cell satisfies the following condition

$$\begin{aligned}
\nu &\leq 1/2\Delta t_{\min} \\
&= \Delta E_{\text{bit}}/2\hbar.
\end{aligned} \tag{3.2}$$

Density

A similar argument may be applied to the minimum size of the cell. Assume that the different logical states are associated with the occupation or not of a certain region of size Δq by the information carrier. Localizing the carrier within Δq imparts to its momentum p an uncertainty Δp given by the position-momentum uncertainty principle:

$$\Delta p \geq \hbar/\Delta q.$$

From $p \geq \Delta p$ the ground-state kinetic energy $E_{\text{kin}|\min}$ of the information carrier satisfies the following inequality chain:

$$\begin{aligned}
E_{\text{kin}|\min} &= p^2/2m \\
&\geq (\Delta p)^2/2m \\
&\geq \hbar^2/2m(\Delta q)^2.
\end{aligned} \tag{3.3}$$

The condition for localization to be possible is

$$E_{\text{kin}|\text{min}} \leq \Delta E_{\text{bit}},$$

so that combining this inequality with (3.3) one has $\Delta q \geq \Delta q_{\text{min}}$, with

$$\Delta q_{\text{min}} = \hbar/\sqrt{2m\Delta E_{\text{bit}}}. \tag{3.4}$$

3.3.2 Limits Imposed by Ballistic Material Motion

A close inspection of the derivation of (3.2) shows that the transition from one state to another has been assumed not to involve any material motion. This assumption, however, is not physical: Whichever process one can hypothesize for modifying the information (redox process, conformational change, change of electronic state, etc.), it is inherently associated with some material motion (redox processes imply electron transfer; conformational or electronic-state changes imply nuclear motions, etc.) in a region which defines the cell size. For that, one has to consider the minimum distance ℓ that the information carrier must travel to allow the cell to switch from one state to another (ℓ may be thought of as the distance separating the minima defining the memory cell and it is expected to be comparable with ℓ_{cell}). If one wants the switching to be externally controlled, tunnel transitions from one state to another must be negligible,[6] so that one can limit the attention to motions with energy higher than ΔE_{bit}. Describing them classically, the carrier average velocity $\bar{v}$ is given by $\bar{v} = \chi\sqrt{2\Delta E_{\text{bit}}/m}$, where χ is a quantity depending on the actual energy of the carrier and on the potential energy inside the cell. Having assumed that the carrier attains its equilibrium configuration in the second cell via dissipation of its kinetic energy, this implies the dissipation of the maximum kinetic energy $(1/2)mv_{\text{max}}^2$ per switching. If one wants to minimize this energy, it must be just above ΔE_{bit}, that gives $\chi \simeq 1/2$.

Assuming that the time required for the ballistic motion, $\ell/\bar{v}$, combines quadratically with the time involved by the uncertainty principle, one has

$$\Delta t \geq \sqrt{(\Delta t_{\text{min}})^2 + (\ell/\bar{v})^2}$$

$$= \frac{\hbar}{\Delta E_{\text{bit}}}\sqrt{1 + \frac{s}{4\chi^2}\left(\frac{\ell}{a_0}\right)^2}, \tag{3.5}$$

---

[6] The effect of the thermal reservoir is considered in the next section. The situation in which the evolution of the system is controlled by the thermal reservoir (rather than by external fields) is considered in Sect. 3.4.

where

$$s = \frac{m}{m_0} \frac{\Delta E_{\text{bit}}}{E_0},\tag{3.6}$$

and length, mass, and energy have been expressed in terms of their natural units a_0, m_0, and $2E_0$ in atomic physics ($a_0 = 1$ bohr $= 0.53\,\text{Å}$, $m_0 = 9.1 \times 10^{-28}$ g – the electron mass – and $E_0 = 1$ rydberg $= 13.6\,\text{eV}$).

Assuming similarly that Δq_{min} combines quadratically with ℓ, one has

$$\begin{aligned}
\Delta q &\geq \sqrt{(\Delta q_{\text{min}})^2 + \ell^2} \\
&= \sqrt{\frac{\hbar^2}{2m\Delta E_{\text{bit}}} + \ell^2} \\
&= a_0 \sqrt{\frac{1}{s} + \left(\frac{\ell}{a_0}\right)^2}.
\end{aligned}\tag{3.7}$$

There is no special reason for assuming quadratic combinations; other (equally arbitrary) choices, like linear combinations, would however produce nearly the same result.

3.3.3 *Limits Imposed by the Thermal Embedding*

The thermal embedding affects the computation in several instances. The following argument, although extremely simplified, provides an idea of the effect of the thermal reservoir on computation. If the particle does really feel the thermal environment, its ground-state energy and energy dispersion must be of the order of $k_{\text{B}}T$ because otherwise they would be thermalized through the collisions with the environment.[7] The uncertainty principle for position and momentum (3.3) would thus limit the best possible localization to a size of width

$$\begin{aligned}
\Delta q_{\text{th}} &\geq \frac{\hbar}{\sqrt{2mk_{\text{B}}T}} \\
&= a_0 \sqrt{\frac{m_0}{m} \frac{E_0}{k_{\text{B}}T}}.
\end{aligned}\tag{3.8}$$

---

[7] Actually, information can be stored, sensed, and modified without ballistic motion by coding the 0 and 1 states by means of energy eigenstates. An excited state embedded in a thermal reservoir, however, decays spontaneously via interaction with the thermal reservoir. This decay may be avoided only in the absence of any coupling – that may be achieved, for instance, exciting, sensing, and de-exciting internal degrees of freedom of nuclei. Since the energy involved in these processes is higher than the thermal energy by several orders of magnitude, its thermalization is highly improbable. Although this route seems in principle interesting for nondissipative computation, in the absence of any idea for its exploitation and in line with the conservative approach of this chapter, it will be ignored.

For $m = m_0$ and $T = 300\,\text{K}$, (3.8) would give to Δq_{th} a minimum value of $1.2\,\text{nm}$. On the other side, the erratic collision with the thermal reservoir while limiting the space localization is ultimately responsible for the decoherence of the quantum state and to its reduction to a classical state; Δq_{th} defines the border between classical and quantum physics: whereas all phenomena occurring on a length scale below Δq_{th} require a quantum description, phenomena occurring on a larger length scale are adequately described by classical physics.

Thus if the energy of the information carrier is controlled by the thermal reservoir, genuine quantum mechanical phenomena (like the Aharonov–Bohm effect, quantum size effect, or tunneling) are manifested on the subnanometer length scale only. Although this length is manifestly short, it however does not preclude the exploitation of phenomena of potential interest (e.g., interference effects for conduction along the side of the benzene ring or of a few condensed aromatic rings [60]).

The detailed prediction of the decoherence length requires complicated theories [61, 62]; it is nonetheless quite intuitive that the decoherence length increases with the stiffness of the material medium hosting the information carrier; the use of materials with high Debye temperature or Einstein oscillator frequency is thus expected to facilitate the appearance and exploitation of such phenomena.

If the information carrier does really undergo thermalization, it can change state, from 0 to 1 or vice versa, via thermal activation. The time τ_T required to migrate from one well to the neighboring one is indeed given by

$$\tau_T = \tau_\infty \exp\left(\frac{\Delta E_{\text{bit}}}{k_\text{B} T}\right). \tag{3.9}$$

with τ_∞ a characteristic period in the ground state. Assume τ_∞ in the interval 10^{-14}–$10^{-12}\,\text{s}$ (corresponding to electronic motion in weakly bound states in the lower regime, or to hindered atomic vibrations in the higher regime); at $300\,\text{K}$ values of ΔE_{bit} around $0.8\,\text{eV}$ would produce lifetimes consistent with dynamic memories (for which the information must be continuously refreshed) whereas ΔE_{bit} around or greater than $1.3\,\text{eV}$ would produce lifetimes consistent with nonvolatile memories:

$$\Delta E_{\text{bit}} = 0.8\,\text{eV} \implies \tau_T = 2 \times 10^{-1}\text{–}2 \times 10^{1}\,\text{s},$$

$$\Delta E_{\text{bit}} = 1.3\,\text{eV} \implies \tau_T = 5 \times 10^{7}\text{–}5 \times 10^{9}\,\text{s}.$$

Equation (3.8) determines not only the best carrier localization but also the minimum distance ℓ separating the regions of the cell coding states 0 and 1:

$$\ell \geq 2\Delta q_{\text{th}}$$

$$= a_0 \sqrt{\frac{m_0}{m}\frac{E_0}{k_\text{B} T}}. \tag{3.10}$$

In a planar arrangement the maximum bit density ρ per unit area cannot be larger than $1/\ell^2$; even more, the consideration that wires and contacts are required to allow the cell to be powered, addressed, and sensed leads to the introduction of a packing factor f_{pack} ($f_{\text{pack}} < 1$) specifying the fraction of space linearly occupied by the cell along the x or y direction:

$$\rho \le (f_{\text{pack}}^2/\ell)^2.$$

From (3.5) and (3.7) one eventually has

$$\rho \le \frac{1}{a_0^2} \frac{f_{\text{pack}}^2 s}{1 + s(\ell/a_0)^2} \tag{3.11}$$

$$\nu \le \frac{\Delta E_{\text{bit}}}{2\hbar} \frac{1}{\sqrt{1 + \frac{1}{4\chi^2} s(\ell/a_0)^2}} \tag{3.12}$$

$$\dot{F} \le \frac{\Delta E_{\text{bit}}}{2\hbar a_0^2} \frac{f_{\text{pack}}^2 s}{[1 + s(\ell/a_0)^2]\sqrt{1 + \frac{1}{4\chi^2} s(\ell/a_0)^2}}. \tag{3.13}$$

If each cycle involves the dissipation of an energy $2\Delta E_{\text{bit}}$, the maximum dissipated power per unit area is given by

$$\dot{W} \le \frac{(\Delta E_{\text{bit}})^2}{\hbar a_0^2} \frac{f_{\text{pack}}^2 s}{[1 + s(\ell/a_0)^2]\sqrt{1 + \frac{1}{4\chi^2} s(\ell/a_0)^2}}. \tag{3.14}$$

For electrons ($m = m_0$) (3.8) and (3.10) give $\ell \le 2.4\,\text{nm}$. Of course, in view of the arguments used for its estimate, this value must be considered only as an order-of-magnitude estimate. In spite of that, since the forthcoming estimates require a knowledge of ℓ, the above value will be used.

Taking $\chi = 0.5$, $f_{\text{pack}} = 0.5$ (a very aggressive packing factor), and $\Delta E_{\text{bit}} = 1\,\text{eV}$ and assuming moreover that the maximum rate is limited by the switching of each cell from one state to another only, the maximum bit density, flop rate, and dissipated power are given by

$$\rho \le 4.3 \times 10^{12}\,\text{cm}^{-2},$$
$$\dot{F} \le 1.7 \times 10^{26}\,\text{cm}^{-2}\text{s}^{-1},$$
$$\dot{W} \le 5.4 \times 10^{14}\,\text{erg}\,\text{cm}^{-2}\,\text{s}^{-1}.$$

The upper value of dissipated power ($5.4 \times 10^7\,\text{W}\,\text{cm}^{-2}$!) is manifestly inconsistent with the properties of matter. Let $\dot{W}_{\text{max}}$ be the maximum allowed power dissipation. The constraint $\dot{W} \le \dot{W}_{\text{max}}$ may thus be satisfied limiting the flop rate to a value lower than $\dot{W}_{\text{max}}/2\Delta E_{\text{bit}}$. For $\Delta E_{\text{bit}} = 1\,\text{eV}$ and $\dot{W}_{\text{max}} = 1\,\text{W}\,\text{cm}^{-2}$ (a value that may assumed to be representative for silicon at $300\,\text{K}$) this would limit the maximum flop rate to $3 \times 10^{18}\,\text{cm}^{-2}\,\text{s}^{-1}$ and determines a maximum parallelism of 8×10^4.

The upper limit of ρ, instead, is not dramatically far from what has already been achieved; suggesting a road for the achievement of this limit within the current technology is one major aim of this book.

3.4 Computation and Irreversibility

The increase of transistor density and speed in ICs has been possible only thanks to the corresponding decrease of energy dissipation per logic operation, the logarithmic rate of decrease having been of 2.5 orders of magnitude in 10 years (from about 10^5 pJ in the early 1960s to about 1 pJ at the end of the 1970s [46]).

There is a large consensus that *the* ultimate physical limit to integration is power dissipation[8] [42, 50–52]. In this line of thought, it is interesting to understand which is the minimum dissipation allowing computation.

3.4.1 Irreversible Computation

Assume first of all that computation is performed synthesizing the computing circuits by means of binary gates (AND, NAND, OR, NOR, XOR, ...) and consider, for instance, the digital output y of a NAND gate (with truth table given in Fig. 3.2) as a function of its two digital inputs x_1 and x_2:

$$y = x_1 \text{ NAND } x_2.$$

Actually, provided that fan-out is available (i.e., that the same output signal can be fed as an input to more that one gate) any binary operation can be synthesized combining NAND gates; the NAND gate is accordingly said to be a universal logic primitive [63].

In view of the truth table of the NAND operation (see Fig. 3.2), if at a certain time the system is in a state with $y = 0$, its inputs are given by $x_1 = x_2 = 1$.

Imagine now that at a later time during the computation y switches from 0 to 1. In the absence of other changes in the state of the system, this transition is manifestly irreversible because it is impossible to recover from the output, the possible inputs ($x_1 = 0, x_2 = 1; x_1 = 1, x_2 = 0;$ or $x_1 = 0, x_2 = 0$) that determine the actual state of the system. A similar argument holds true if the synthesis of the circuit was done employing the NOR gate (the truth table is given in Fig. 3.2) instead of the NAND gate. In view of the irreversibility of the above process (nothing but a little step of computation) *the entire computation act is an irreversible process* too. In this way, the free energy lost in each switch is wasted as dissipated heat. Since in general the

---

[8] It is worthwhile noticing that the shift of technology, from n-channel MOS to CMOS, was not motivated by a corresponding increase of density, but rather by a substantial decrease of power dissipation allowed by the new technology.

operation	truth table	algebraic notation	symbol

NOT

x	y
0	1
1	0

$y = \overline{x}$

AND

x_1	x_2	y
0	0	0
0	1	0
1	0	0
1	1	1

$y = x_1 \cdot x_2$

OR

x_1	x_2	y
0	0	0
0	1	1
1	0	1
1	1	1

$y = x_1 + x_2$

NAND

x_1	x_2	y
0	0	1
0	1	1
1	0	1
1	1	0

$y = \overline{x_1 \cdot x_2}$

NOR

x_1	x_2	y
0	0	1
0	1	0
1	0	0
1	1	0

$y = \overline{x_1 + x_2}$

XOR

x_1	x_2	y
0	0	0
0	1	1
1	0	1
1	1	0

$y = x_1 \oplus x_2$

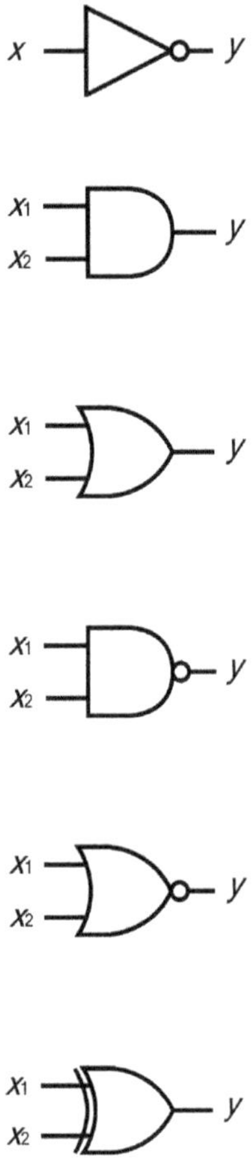

Fig. 3.2 Truth tables, algebraic notations, and symbols for NOT, AND, OR, NAND, NOR, and XOR operations

number of switches is very much larger than the number of outputs, this occurrence generates the problem of minimizing the energy dissipation per switching event.

3.4.2 Reversible Computation

The argument showing the irreversibility of computation fails if all states generated during the computation are memorized. The practical exploitation of this observation is limited by two factors:

- The large (usually very large) number of states generated during the computation would require computers with correspondingly large memories.
- It is not clear if the reset of the auxiliary memory is or is not a dissipative process.

A radical approach, avoiding the use of auxiliary memories and their reset, is based on the use of invertible gates. They rest on the idea that the energy carried by the signal should not be dissipated and that any gate must provide an output with the same energy content as the input. Thus, contrary to what happens in ordinary logic, any output cannot be used to feed more than one logic gate (fan-out is 1) because information is considered as a kind of resource [64]; thus, signal copies cannot be taken by tapping a wire, but rather are transmitted by means of gates; in so doing, they are (temporarily) created only when they are needed. Of course, in so doing computation requires, in addition to the argument, constant input and produce, in addition to the result, unrequested outputs. Actually Fredkin and Toffoli have invented such gates and shown that they are universal gates [44][9]:

the **Fredkin gate** has three inputs (u, x_1, x_2) and three outputs (v, y_1, y_2). The first signal, u, always goes unchanged through the gate, while the other two come out either straight or swapped depending on whether u equals 1 or 0:

$$v = u,$$
$$y_1 = ux_1 + \overline{u}x_2,$$
$$y_2 = \overline{u}x_1 + ux_2.$$

the **Toffoli gate** has three inputs (u, x_1, x_2) and outputs (v, y_1, y_2) satisfying the following relationships:

$$y_1 = x_1,$$
$$y_2 = x_2,$$
$$v = u \oplus (x_1 x_2).$$

For both Fredkin and Toffoli gates the entire energy of the output signals comes from the input signals.

Fredkin and Toffoli did not limit themselves to the description of abstract conservative gates, but also found a mechanical model (classical billiard) for them [44]. Starting from the idea of modeling a gate as something mimicking an elastic collision, Toffoli also proposed the use of charges for computations [63].

Aside as it was from the main stream of electronics and in spite of recent excitement about molecular cellular automata [65], this proposal remained however without practical application and the debate has concentrated on the possibility of

---

[9] In the following, algebraic symbols are used to denote logical operations: overlining is used for negation, simple juxtaposition denotes AND, "+" denotes OR, and "$\oplus$" denotes XOR; thus "$\overline{u}x_1 + ux_2$" means "(NOT(u) AND x_1) OR (u AND x_2)" whereas "$u \oplus (x_1 x_2)$" means "u XOR (x_1 AND x_2)."

accepting some degree of dissipation (because computing devices operate in a world dominated by the second law of thermodynamics), trying to exploit it (Brownian computers [45]) or at least to minimize it.

Whereas the practical exploitation of Brownian motion will be considered in Chap. 11, the next section is devoted to understand to which limits the dissipated power can be reduced in ultimate Brownian computers.

3.4.3 Minimum Dissipation

The model of memory so far considered will be slightly modified, however retaining the basic idea that each cell, immersed in a heat reservoir at temperature T, is formed by two rooms numbered 0 (say the one on the bottom in Fig. 3.3) and 1 (the one on the top) of equal volume ω, in which the information carrier is localized. Assume initially that the information carrier is an ideal gas of N_g molecules at a pressure P of the order of one atmosphere. Such a gas obeys the ideal gas law:

$$PV = N_g k_B T, \tag{3.15}$$

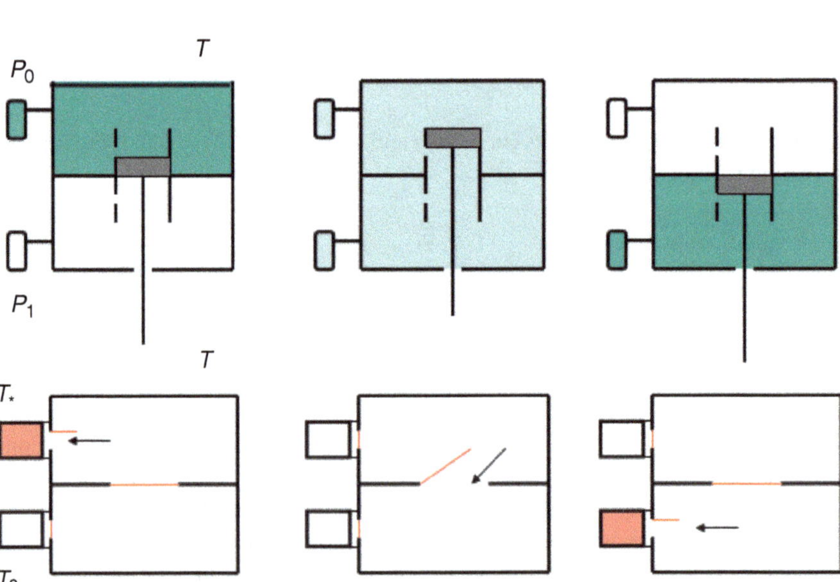

Fig. 3.3 Schemata of ideal memory cells exploiting a dense gas (*top*) or a single molecule (*bottom*) as information carrier. *Top*: the gas is compressed by means of a frictionless piston and a suitable combination of closure and opening of frictionless valves (not shown) into room 1 (*left*); the transition from 1 to 0 is achieved opening the orifice separating the rooms (*middle*) and compressing the gas into room 0 (*right*). The state of the system is detected by manometers. *Bottom*: after enabling of detector 1, the frictionless door separating the rooms is left open until the molecule is detected (*left*); the transition from 1 to 0 is carried out enabling detector 0 and opening the door (*middle*) until the detector feels the particle (*right*); no work is done but any change of state (of which one has information only after a measurement) requires energy

where V is the actual volume where the gas is contained. State 0 is characterized by the condition that the gas is in room 0 whereas state 1 is characterized by the gas in room 1. Starting from an initial condition where the gas is spread over all the volume 2ω ($V = 2\omega$) one can prepare the system in state 1 (or 0) compressing all the gas in the top (or bottom) room ($V = \omega$). If this process is carried out under near equilibrium isothermal conditions, the process requires an external work given by

$$L = -\int_{2\omega}^{\omega} P \, dV$$
$$= N_g k_B T \ln(2), \tag{3.16}$$

where (3.15) was used. For a perfect gas the internal energy U does depend only on T, so that U does not vary in the compression and work (3.16) is released as heat Q to the heat reservoir; if this process is carried out reversibly, it results in a change (actually a decrease) ΔS of the entropy of the system by an amount given by $T\Delta S = L$. The insertion of (3.16) gives

$$\Delta S = -N_g k_B \ln(2). \tag{3.17}$$

Refer now to the above quantities, L and ΔS (as given by (3.16) and (3.17)), to each molecule forming the gas: one has

$$L/N_g = k_B T \ln(2), \tag{3.18}$$
$$\Delta S/N_g = -k_B \ln(2), \tag{3.19}$$

and consider what happens when N_g is progressively reduced to 1.

This matter, of course, is strongly related to the problem of the Maxwell demon. Although the Maxwell demon appeared in physics in 1867 (in a letter to Tait) and was revived in 1875 in Maxwell's *Theory of Heat* [66], and in spite of the analyses of Szilard [67] and Brillouin [68], it has not been fully exorcized yet. In fact, the amount of work that the demon can extract from a system at constant temperature as well as the intriguing links between macroscopic irreversibility and microscopic reversibility are still a matter of debate [69–76].

In the progressive reduction of N_g, three regimes can be considered:

- At high pressure the fluctuations in the number N_{coll} of molecular collisions on the manometer producing what is externally detected as an internal pressure are totally negligible, the involved quantities are macroscopic, and all phenomena are described in thermodynamic terms.
- At sufficiently low pressures such fluctuations are not negligible; if the impacts on the membrane are described by a Poisson process, the fluctuations vary as $\sqrt{N_{\text{coll}}}$.
- At extremely low pressures the membrane feels a collision with gas-phase molecules only occasionally while it remains unperturbed for most time.

The three regimes are thus determined by the value assumed by N_{coll}:

$$N_{\text{coll}} \begin{cases} \gg 1 : \text{absence of fluctuations,} \\ \approx 1 : \text{domination of fluctuations,} \\ \ll 1 : \text{domination of single events.} \end{cases}$$

The quantity N_{coll} depends on instrumental factors (the area A of the manometer membrane and the time constant τ over which the average displacement of the membrane is taken) and on the flux Φ of molecules impacting the membrane[10]:

$$N_{\text{coll}} = \Phi A \tau.$$

According to the kinetic theory of gases, $\Phi = P/\sqrt{2\pi m k_B T}$ so that the above regimes are controlled by the quantity $(P/\sqrt{2\pi m k_B T})A\tau$. To give a quantitative idea, observe that for $A \approx 1\,\mu\text{m}^2$, $\tau \approx 10^{-3}\,\text{s}$, $m \approx 5 \times 10^{-23}\,\text{g}$ (air gases), and room temperature, one has $N_{\text{coll}} \approx 1$ for $P \approx 4 \times 10^{-7}\,\text{dyn cm}^{-2}$ ($\simeq 3 \times 10^{-10}\,\text{Torr}$). Consider now what happens when the system is formed by one molecule only. Although a system formed by a single molecule in a given volume can hardly be thought of as an ideal gas, (3.19) continues to hold true provided that the entropy is considered in a statistical sense: in fact, reducing the volume by a factor of 2 implies that the number of available states is reduced by the same amount so that the Boltzmann formula of entropy yields exactly (3.19).

More difficult is to assess anything about the minimum work. In fact, one can speak of pressure only for macroscopic systems containing so many particles that their collisions against the walls are indeed collectively manifested as a pressure. Obviously, this is not the case when the information is encoded in one particle only.

In this case, the experiment cannot be done pumping the gas from one room to the other, but something similar is obtained modifying the experimental setup as follows: The system is formed by two rooms, each endowed with a cryogenic calorimeter, a device able to detect whether the particle (otherwise freely moving between collisions on the walls) is there. The rooms are separated by a door that can be opened or closed at will without friction. Something similar to this arrangement is sketched in Fig. 3.3.

The preparation of the system starting from an initial state for which it is not known, where the particle resides (open door), to a state coding 1 (particle in room 1) is achieved via the following sequence of events:

---

[10] It is common practice in vacuum physics to define *exposure*, the product $\Phi\tau$, where τ is the duration over which the flux is active, and to use the langmuir as the unit of exposure:

$$1\,\text{langmuir} = 10^{-6}\,\text{Torr s,}$$

where the Torr is a unit of pressure of historical and practical relevance (together with the atmosphere, $1\,\text{atm} = 760\,\text{Torr}$) not included in the SI. In SI units $1\,\text{Torr} = 1.33 \times 10^2\,\text{Pa}$. Although extraneous to any accepted unit system, the langmuir is a useful unit: an exposure to 1 langmuir of air at room temperature corresponds to $3.6 \times 10^{14}\,\text{cm}^{-2}$ – roughly one monolayer.

P_1, enabling the detector in room 1
P_2, waiting until the detector reveals the particle
P_3, closing the door without friction
P_4, disabling the detector in room 0

The time required by this operation is controlled by the ratio of the cross section of the detector to the total area of inner walls; the operation is subjected to an error that increases with the delay between the detection of the particle and the actual closure of the door and with the ratio of the area of the door to the area of inner walls of room 1.

This situation, that reiterated for a set of molecules would result in the accumulation in one room of a gas originally in two rooms, is similar to that of a Maxwell demon when he (or she) tries to separate fast from slow molecules. According to the analysis of Szilard [67] and Brillouin [68], this operation requires feeding the detector with a minimal external energy, given by (3.18), sufficient to balance the decrease of entropy.

After a few collisions on the room walls, the molecule will reacquire a kinetic energy of about $(3/2)k_B T$; once that has been achieved, the state of the system can be determined enabling both calorimeters and waiting until one or the other feels the particle; the energy involved in this operation is the same as that required for the preparation of the system.

State 0 and state 1 are characterized by the same entropy, so that one might argue that switching does not require any dissipation [73]. This point is critical because it would suggest that the reset of the auxiliary memory considered in the previous section does not require any energy loss. However, resting to the ideal situation considered above, consider which sequence of events allows the transition from state 1 to state 0:

S_1, opening (without friction) the door
S_2, enabling the detector in room 0
S_3, waiting until the detector reveals the particle in room 0
S_4, closing (without friction) the door
S_5, disabling the detector

A comparison between sequences P_1–P_4 and S_1–S_4 shows that in each of them there is only a dissipative process (associated with the detection of the particle) so that one may conclude that the minimum energy dissipation required for switching in a thermal reservoir at temperature T, where the computing device is embedded, is obtained assuming that the minimum energy dissipation ΔE_{min} is determined by the change of configurational entropy ΔS:

$$\Delta E_{min} \simeq -T\Delta S \tag{3.20}$$

$$= k_B T \ln(2). \tag{3.21}$$

This formula, first asserted (without justification) by von Neumann and confirmed by the Landauer analysis [40], is supported for more realistic arrangements by the discussion of Meindl and Davis [50].

This long discussion shows an example of transition from one state to another (both with the same a priori probability) occurring through a state where the particle is statistically dispersed in both states. The analysis also seems to suggest that the energy dissipation is proportional to the change of entropy resulting from the contraction of the volume.

In these terms, the conclusion is false and requires a better specification.

Consider indeed what happens when the rooms have different volumes (say $\alpha\omega$ and $(2-\alpha)\omega$, with $0 < \alpha < 2$). The change of configurational entropy resulting from the localization first in one room and then in the other is thus given by $k_B \ln(\alpha/2) + k_B \ln((2-\alpha)/2)$, so that on the average the entropy change ΔS_α is given by

$$\Delta S_\alpha = \tfrac{1}{2} k_B \ln\left(\frac{\alpha(2-\alpha)}{4}\right). \tag{3.22}$$

Since ΔS_α depends on α whereas the energy ΔS_{bit} involved in the calorimetric detection does not, one could conclude that the tentative interpretation above is not correct. Rather, observing from (3.22) that ΔS_α has a maximum for $\alpha = 1$ and that $\Delta S_{\text{max}} = \Delta S_1 = k_B \ln(2)$, *(3.21) continues to hold true provided that the change of entropy therein coincides with the maximum change of entropy ΔS_{max} involved in the localization in one state between two assigned ones.*

This analysis allows a generalization: fixing one logical state among n assigned states requires a minimum energy given by

$$\Delta E_{\text{min}} = k_B T \ln(n). \tag{3.23}$$

3.4.4 Computation and Measure

The above discussion shows that computation is ultimately related to measurement. On another side, measurements are considered in very different fashions in classical and quantum physics. The following discussion will be limited by the assumption that the information carrier has a classical nature.

Classical Measurement

In the *Gedankenexperiment* considered in the following the detection of the information carrier plays a role so important to render it necessary that its specification is given with more details.

Perhaps the conceptually simplest detector is the calorimeter. In principle, the particle may be detected by measuring the change of temperature resulting after the collision of the molecule on a calorimeter.[11] Obviously, the calorimeter must be very sensitive and be at a temperature T_{cal} different from that of the thermal reservoir (otherwise, the collisions would on the average occur without exchange of energy). Let Q be the energy imparted to (or released from) the calorimeter during the collision. If $C_{V,cal}$ denotes the heat capacity at constant volume of the calorimeter, the change of temperature ΔT produced by the collision is given by

$$\Delta T = Q/C_{V,cal}. \tag{3.24}$$

On another side, the internal energy of the calorimeter fluctuates by an amount ΔU_{cal} given by [77]

$$\Delta U_{cal} = \sqrt{k_B C_{V,cal} T_{cal}^2}. \tag{3.25}$$

Due to those fluctuations, the temperature change (3.24) is actually detectable only if $|Q| > \Delta U_{cal}$.

The efficiency of the detection is maximized taking $|Q|$ as large as possible and ΔU_{cal} as small as possible. Since the heat capacity is an increasing function of temperature, the second condition is better satisfied working at $T_{cal} < T$, in which case $Q > 0$.

The maximum efficiency is achieved for Q as close as possible to the total kinetic energy $(3/2)k_B T$ of the particle. It is however noted that if one wants that the particle is not adsorbed by the calorimeter (thus leaving the subsystem where it was trapped) only a fraction of $(3/2)k_B T$ may be imparted to the calorimeter (that therefore must be viewed as an *imperfect calorimeter*); say $Q \simeq k_B T$. Combining this argument with (3.25) gives

$$T_{cal}/T \lesssim \sqrt{k_B/C_{V,cal}}. \tag{3.26}$$

On another side, the calorimeter is a macroscopic device, formed by a large (compared to 1) number N_{cal} of atoms. Since $C_{V,cal} \propto N_{cal}$ and vanishes for $T_{cal} \to 0\,K$, T_{cal} must be much lower than T. The closer T_{cal} to T, the closer the overall system (rooms and calorimeter) to equilibrium and the lesser the detection is energetically expensive. Trying to operate with T_{cal} as close as possible to T requires that the calorimeter has its minimum size.

The minimum number of atoms required for defining a calorimeter may be estimated following closely the analysis of the thermal spikes resulting from ion implantation as given by Cerofolini and coworkers [78–80]. The existence of a minimum size of the calorimeter derives from the need of transforming all the energy imparted by the collision into heat. In fact, in the early stages after the impact only few atoms share the imparted energy while the remainder are in equilibrium at temperature T_{cal}. Assuming that the energy is dispersed via two-body atomic

---

[11] A calorimeter is a device able to absorb the entire kinetic energy of particle transforming it into heat.

collisions, in the early stage of its formation the collisional cascade involves 2^κ atoms, where κ is the branching order of the cascade; if the atoms in the cascade have low velocity, their mean free path is of the order of the interatomic distance so that when 2^κ exceeds the number of atoms contained in a sphere of radius κ, $(4/3)\pi\kappa^3$, the atoms collide with atoms already in motion and the initial energy is shared by all atoms of the cascade. The condition $2^\kappa = (4/3)\pi\kappa^3$ identifies a critical branching, $\overline{\kappa} = 13.25$, and a cascade size, $(4/3)\pi\overline{\kappa}^3 \simeq 10^4$, determining the minimum calorimeter size,

$$\overline{N}_{\text{cal}} \simeq 10^4, \tag{3.27}$$

for which the cascade can be seen as a collection of atoms at a temperature T_{cal} different from the ambient one.

For the present purposes, the thermal properties of crystalline solids are adequately described by the Debye theory. In this theory the heat capacity $C_V(T)$ of any crystal is a universal function of its normalized temperature T/T_{D}, with T_{D} the Debye temperature. This function has the following asymptotic behavior:

$$C_V(T) \simeq \begin{cases} 3Nk_{\text{B}} & \text{for } T/T_{\text{D}} \gg 1 \text{ (Dulong–Petit law)}, \\ \frac{12\pi^4}{5}Nk_{\text{B}}(T/T_{\text{D}})^3 & \text{for } T/T_{\text{D}} \ll 1, \end{cases} \tag{3.28}$$

where N is the number of the atoms forming the solid.

Assume now that the calorimeter is a crystalline solid with $N = \overline{N}_{\text{cal}}$ and the imparted energy is totally spent to excite its thermal vibrations. The maximum operating temperature of the calorimeter is obtained specializing (3.26) to the present case. If the Debye temperature of the calorimeter is low, its heat capacity is described by the Dulong–Petit law, and (3.25) gives

$$T_{\text{cal}} < (3\overline{N}_{\text{cal}})^{-1/2}T$$
$$\simeq 6 \times 10^{-3}T. \tag{3.29}$$

For $T = 300$ K (3.29) states that the detection with the smallest possible calorimeter is possible only in the cryogenic regime, say $T_{\text{cal}} \lesssim 1$ K.

The situation changes for solids with Debye temperature sufficiently high to allow them to run in the second case of (3.28):

$$T_{\text{cal}} < \left(\frac{T_{\text{D,cal}}^3 T^2}{\frac{12\pi^4}{5}\overline{N}_{\text{cal}}}\right)^{1/5}. \tag{3.30}$$

To give an idea of the limits imposed by (3.30), consider the use of diamond or silicon calorimeters. Using diamond, the substance with the highest Debye temperature ($T_{\text{D}} = 2220$ K) [81], one should operate at $T_{\text{cal}} \lesssim 50$ K. For silicon calorimeters ($T_{\text{D}} = 645$ K), (3.30) gives instead $T_{\text{cal}} \lesssim 25$ K.

The cryogenic detection of rare events has been an intense field of research in last years especially for the detection of nuclear or cosmic events [82–85]. Because of this, in most cases the attention was concentrated on large calorimeters, for which the operating temperature must be in the deep cryogenic regime.

The sensitivity required for the detection of a single molecule at room temperature ($k_B T = 25\,\text{meV}$) is not yet at the reach of current possibilities; however, it is not the only factor. Also vacuum technology poses problems: Just to give an idea of the orders of magnitude, in currently achievable ultra-high vacua the residual pressure is of the order of 10^{-10} Torr ($\simeq 10^{-7}\,\text{dyn}\,\text{cm}^{-2}$). A volume of $1\,\text{cm}^3$ kept at room temperature contains approximately 3×10^6 molecules. At this pressure each surface atom of the calorimeter undergoes a collision against gas-phase molecules each 10^4 s and the entire surface exposed to the atmosphere of the minimum-size calorimeter, of about 10^3 atoms, feels an event each 10 s.

Quantum Measurement

The limits following from the analysis of Sect. 3.4.3 are essentially due to the fact that the localization of the information carrier in one or the other of its states is the result of a measurement process, albeit classical. This might be strange were it not for the fact that both classical and quantum measurements perturb the state of the particle; what renders them different is the effect of this perturbation: predictable in classical physics, unpredictable in quantum physics.

Whereas the situation admits a simple rationale in classical terms, it is much less clear when the information carrier is assumed to have a quantum nature. What one can certainly assert is that since the eigenstates $|0\rangle$ (particle in room 0) and $|1\rangle$ (particle in room 1) are orthogonal, the transition from one state to another requires a minimum time given by $\hbar/\Delta E$, with ΔE the energy dispersion of each localized state. This time is long enough to make observable the state where the particle is spread over *both* rooms so that the localization in the other room requires a negentropy eventually responsible for (3.21). It seems thus reasonable to take (3.21) as an estimate of the minimum energy involved in any detectable switch.

In view of (3.14) any reduction of ΔE_{bit} would result in a quadratic reduction of dissipated power; thus reducing ΔE_{bit} from 1 eV to 20 meV would result in a reduction of the dissipated power by a factor of about 2×10^3. At this stage, however, no strategy has yet been identified for the design and preparation of near-to-equilibrium logic devices.

3.5 Reading

As far as decoherence destroys the quantum nature of the state (transforming the quantum logic states $|0\rangle$ and $|1\rangle$ into classical logic states 0 and 1) the determination of the memory cell state does not involve the characteristic problems associated with the measurement of the quantum system.

3.5.1 Coupling the Carrier with the External World

The classical nature of the information carrier allows the specification of the most convenient quantity for its management.

Excluding spin (because it is intrinsically a quantum property disappearing in the classical limit) the attention is focussed on the material properties which characterize matter: mass, charge, electric dipole, magnetic dipole, refractive index, etc. Of them, the most convenient choice is thus dictated by the ease of actuating the $0 \rightleftarrows 1$ transitions (PROGRAM operation) and sensing the corresponding states (READ operation).

Excluding mass (that can be actuated and sensed through gravitational fields, in general extremely weak with respect to the Earth one), one must focus attention on the electromagnetic quantities – and among them the electric charge is the quantity that allows the easiest manipulation.

If the information carrier is charged, the state can be measured by embedding one region of the memory cell (for instance, the one coding state 1) in an electrometer and connecting it to the gate of an MOS transistor. To be a little bit more concrete, assume the electron as information carrier and imagine that in the absence of the electron (electron in the zone coding 0) the transistor is in OFF condition whereas in the presence of the electron the transistor is switched ON.[12] In this arrangement, the macroscopic output will be at ground potential (OUT $= 0$) when the electron is in region coding 0 (responsible for the transistor in OFF condition) or will have a positive voltage (OUT $= 1$) when the electron is in the region coding 1 (responsible for the transistor in ON condition).

Of course, the switching of an MOS transistor requires more than one electron (actually, 10^2–10^4 electrons are currently required for that). However, experiments involving the conduction along molecules have demonstrated that certain molecules are able to store a single electron in a metastable state and that the conductance in this state is higher than that in the neutral state by a measurable amount. This matter is discussed in detail in Chap. 7. The following part is addressed instead to estimate the maximum reading frequency.

3.5.2 Physical Limits in READ *Operation*

As far as the measurement of the conduction state of the cell state is essentially a classical measurement, it can be attacked with elementary considerations. It is particularly simple if the sensing element of the memory cell can be regarded simply as a reconfigurable resistance R^J. This resistance may assume one or the other of

---

[12] That would occur for hypothetical p-channel MOS transistor with gate capacity such that the addition of a unitary charge produces an increase of surface potential able to bring the transistor from accumulation to inversion – about 1 V for silicon.

Fig. 3.4 The cell is read sensing the current flowing through a resistance whose value is controlled by the charge in one half cell

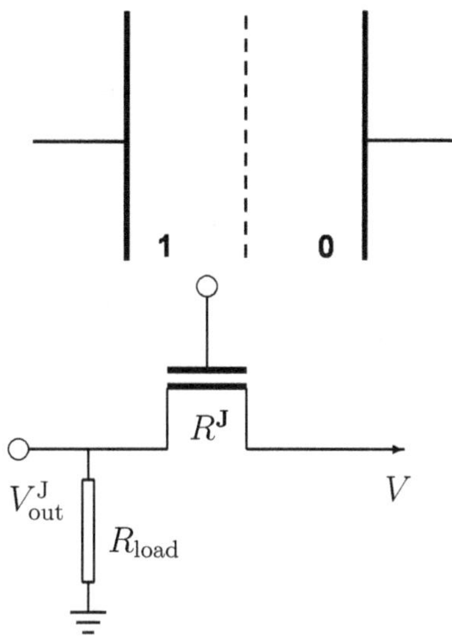

two possible values, $R^J = R^0$ when in OFF condition or $R^J = R^1$ when in ON condition, with $R^1 \ll R^0$.

Denoting, as sketched in Fig. 3.4, the supply voltage with V, the output voltage with V_{out}^{J}, the load resistance with R_{load}, the current flowing through the sensing element with I^J, and the dissipated power with $\dot{W}^J$, one has

$$I^J = \frac{V - V_{\text{out}}^{J}}{R^J}, \tag{3.31}$$

$$V_{\text{out}}^{J} = \frac{R_{\text{load}}}{R^J}(V - V_{\text{out}}^{J}), \tag{3.32}$$

$$\dot{W}^J = \frac{R_{\text{load}} + R^J}{(R^J)^2}(V - V_{\text{out}}^{J})^2. \tag{3.33}$$

The minimum time required for reading the cell state is immediately obtained from (3.31): Whichever is the resistance state J, the electron flux Φ^J flowing through the resistance is given by

$$\Phi^J = (V - V_{\text{out}}^{J})/eR^J. \tag{3.34}$$

with $-e$ being the electron charge.

Assume first that the resistance is in the ON state, $J = 1$. The measurement is actually able to sense the ON state only if the duration lasts a time Δt so long as to allow the passage of at least one electron. This happens when $\Phi^1 \Delta t > 1$, i.e., when

$$\Delta t \gtrsim eR^1/(V - V_{\text{out}}^{1}). \tag{3.35}$$

On another side, the system will detect the passage of an electron even when the molecule is in the OFF state provided that the cell is observed for a time longer than $eR^0/(V - V_{\text{out}}^0)$. As $V_{\text{out}}^0 < V_{\text{out}}^1$, the detection of the passage of one or more electrons in the time interval

$$\{\Delta t : eR^1/(V - V_{\text{out}}^1) \ll \Delta t \ll eR^0/(V - V_{\text{out}}^1)\} \tag{3.36}$$

may be considered to provide evidence for resistance in the ON state; missing any evidence for flowing electrons in time interval (3.36) may instead be considered evidence for resistance in the OFF state. Let

$$\Delta t^{\text{READ}} := eR^1/(V - V_{\text{out}}^1);$$

the quantity Δt^{READ} is thus the minimum time[13] required for the detection of the state, 0 or 1, of the cell.

The time sensitivity can be improved by increasing the potential difference $V - V_{\text{out}}^J$ (of course, maintaining it below the breakdown potential). In so doing, however, the dissipated power increases quadratically because of (3.33).

Since dissipation decreases with the total resistance $(R_{\text{load}} + R^J)$, the attention can be limited to the case of resistance in ON condition; assuming that even reading is limited by power dissipation, $V - V_{\text{out}}^J$ must thus be as close as possible to the minimum voltage involved in the computation:

$$e(V - V_{\text{out}}^1) \simeq k_{\text{B}} T \ln(2). \tag{3.37}$$

Assigning to $V - V_{\text{out}}^1$ its minimum value, $V - V_{\text{out}}^1 = k_{\text{B}} T \ln(2)/e$, (3.33) requires that, for any given R^1, R_{load} is as small as possible. On another side, if one requires that even V_{out}^1 is able of computation (i.e., $V_{\text{out}}^1 \geq k_{\text{B}} T \ln(2)/e$), (3.32) gives $R_{\text{load}} \geq R^1$. Combining this constraint with the need for minimizing dissipation one gets the optimum combination:

$$R_{\text{load}} = R^1. \tag{3.38}$$

The insertion of (3.37) into (3.35) gives

$$\Delta t_{\text{min}}^{\text{READ}} \simeq \frac{e^2 R^1}{k_{\text{B}} T \ln(2)}; \tag{3.39}$$

---

[13] Actually, assuming that the electron flow is a Poisson process, Δt^{READ} is just the time at which the standard deviation of the number of detected electrons coincides with the average number. Statistical significance is obtained only for

$$\Phi^1 \Delta t^{\text{READ}} \gg \sqrt{\Phi^1 \Delta t^{\text{READ}}}.$$

This inequality may be specified quantitatively as $\sqrt{\Phi^1 \Delta t^{\text{READ}}} > r$, with $r \simeq 3$. This condition reads $\Phi^1 \Delta t^{\text{READ}} \gtrsim 10$, i.e., $\Delta t^{\text{READ}} \gtrsim 10 eR^1/(V - V_{\text{out}}^1)$.

the minimum power dissipation in reading a single cell for the optimum circuit is instead obtained by inserting (3.37) into (3.33):

$$\dot{W}_{min} = \frac{2e^2}{R^1 [k_B T \ln(2)]^2}. \qquad (3.40)$$

There remains to discuss R^1, the sensing resistance. It is extremely difficult to assert anything about molecular conductance in general, especially in view of the many mechanisms responsible for the transport through molecules (tunnel, field emission from the contacts, Poole or Poole–Frenkel conduction, etc.) [86]. The discussion is greatly simplified if the attention is focused on carrier conduction occurring from one contact to another through a classically forbidden region. In this case, indeed, one has

$$\frac{1}{R^J} = \frac{e^2}{h} T^J / \left(1 - T^J\right)$$

$$\simeq \frac{e^2}{h} T^J, \qquad (3.41)$$

where $h = 2\pi\hbar$, $h/2e^2$ is the Landauer resistance (a kind of minimum contact resistance associated with any classically forbidden region, $h/2e^2 = 12.9\,\text{k}\Omega$), and T^J (usually $T^J \ll 1$) is a quantity associated with the transparency of the barrier separating the two contacts and with the density of states in the arrival region ((3.41) is discussed in Sect. 7.1.1). This case is particularly interesting because T^J can be modified easily by electric fields in the vicinity of the tunneling region [generated, for instance, by the presence ($J = 1$) or absence ($J = 0$) of an electron in its vicinity – that allows in principle an easy implementation; see Sect. 7.4].

The insertion of (3.41) into (3.39) and (3.40) gives

$$\Delta t_{min}^{READ} \simeq \frac{h}{k_B T \ln(2)} \frac{1}{T^1} \qquad (3.42)$$

and

$$\dot{W}_{min}^J = 4 \frac{[k_B T \ln(2)]^2}{h} T^J. \qquad (3.43)$$

Equations (3.42) and (3.43) specify how Δt_{min}^{READ} and $\dot{W}_{min}^J$ depend on the environment (through T) and on internal properties of the resistance (through T^J).

As already mentioned, it is difficult to assert anything about R^J. In general, for the molecules which have been considered of potential interest, currents in ON condition of the order of 1 nA for applied voltages of 1 V (corresponding to effective resistances of the order of $10^9\,\Omega$, thus much higher than the Landauer resistance, $h/2e^2 = 12.9\,\text{k}\Omega$) have been found, corresponding to $T^1 \approx 10^{-5}$.

Assuming the condition of minimum dissipation, for $T^1 \approx 10^{-5}$ (3.42) and (3.43) give $\Delta t_{min}^{READ} \simeq 2.4 \times 10^{-8}$ s and $\dot{W}_{min}^1 \simeq 4.7 \times 10^{-13}$ W – that would allow the simultaneous reading of all memory cells if they are packed at the maximum density[14] and the allowed dissipation is of the order of $1\,\mathrm{W\,cm^{-2}}$.

3.5.3 A Little Step Toward Practical Implementation

Up to now it has been assumed that the condition $\mathcal{T}^{\mathrm{OFF}} \ll \mathcal{T}^{\mathrm{ON}}$ may be achieved quite irrespective of the applied signal. Actually, the sketch of Fig. 3.4 would suggest that the minimum variation of input potential required to switch the transistor from OFF to ON is that required to bring the transistor from accumulation to inversion – roughly 1 V for silicon (E_{g}/e, with E_{g} the energy gap). For $\mathcal{T}^{\mathrm{ON}} \approx 10^{-5}$ (3.39) and (3.40) give access time and dissipated power per cell of the order of 10^{-9} s and 10^{-10} W, respectively. The last datum shows that in this situation only a part of the total allowed cells could be simultaneously managed to be consistent with an allowed dissipation of $1\,\mathrm{W\,cm^{-2}}$.

An important quantity not considered yet is the current flowing through the reconfigurable resistance in ON: of the order of 1 nA for a supply voltage of 1 V or about 20 pA for a supply voltage of $k_{\mathrm{B}}T \ln(2)/e$. Both these currents are easily measurable, but with an external amplifier only. If one needs to integrate it into the control circuitry (that is mandatory if a number of cells must be managed in parallel) this value would be reduced on the $10\,\mu\mathrm{A}$ current scale. The effect of this technological constraint is considered in Sect. 5.4.

For the present purposes it is sufficient to remark that so far the ways the signals of PROGRAM and READ are imparted to the cell have not been specified yet. Irrespective of the detailed modes used to vary cell potentials or to measure flowing currents, the corresponding involved apparatuses are macroscopic, with a size in general much larger than that of the cell. If these apparatuses must be arranged in a planar configuration too, *only an architecture allowing the sensing of all cells by relatively few measuring apparatuses permits terascale integration*. How such a goal may be achieved in practice will be the matter of Chaps. 4–8.

---

[14] The cell density hitherto considered allows quantum states to be reduced to classical states. Higher densities could be achieved provided that the quantum states are continuously refreshed, thus mimicking the behavior of dynamic memories. That could be achieved via quantum Zeno effect by repeating the measurement of any state in a time much shorter than the time required for the spreading of the wave function to adjacent. This occurs when $2m\ell^2/\hbar \gg e/I$: the current able to stabilize an electron ($m = m_0$) coding a quantum bit from spreading to an adjacent cell separated by a distance ℓ of 1 nm is however higher than $10\,\mu\mathrm{A}$.

Chapter 4
The Crossbar Structure

As already discussed in Sect. 1.3, nanoelectronics (and especially molecular electronics) requires the development of new architectures and of the corresponding appropriate technologies. Both these tasks seem very difficult. The difficulties in the application of molecular electronics for the production of commercial devices are immediately understood comparing the slow development of this discipline to what happened after the invention of the transistor: 5 years after the news became public, one could buy a transistor radio and engineers were experimenting with computer circuits. The reason was that it was easy to build transistors and make them work, and they offered great advantages in terms of power, speed, and reliability. This argument, freely taken from a message sent to the author by Keyes [87], is elaborated in detail in [88].

Building transistors, however, has become, if not difficult, certainly very expensive. On the other side, an old circuit architecture, of difficult use when employing MOS-FETs, has demonstrated potentials of practical application when using molecular devices – the *crossbar structure*. Although particle physicists have become familiar with the crossbar structure as position-sensitive detector (and known in this field as multiwire chamber [89–91]) since the early 1970s,[1] its exploitation in electronics is not trivial. The following features, however, render it of interest for next-generation ICs [93–97]:

- It is formed by two arrays of conducting wires arranged in parallel planes and oriented perpendicularly to one another.
- It may be prepared to include in each crosspoint the memory cell.
- It requires for its construction only definitions (not alignments).
- It is producible (via nonconventional lithography or even without any lithographic method) with geometry on the 10-nm length scale.

---

[1] It is interesting to compare the best spatial resolution offered by multiwire chambers, about $10^2\,\mu$m, with that achievable with sublithographic techniques, of the order of $3 \times 10^{-2}\,\mu$m; it is also noted that the use of conventional lithographic techniques and devices allowed a spatial resolution of the order of $10\,\mu$m [92] (30 years ago).

Although the crossbar is not yet a circuit, it may nonetheless become a circuit if each memory cell can be addressed, written, and read – that requires an external circuitry for addressing, power supplying, and sensing.

Not only does the crossbar architecture allow the preparation of matrices with crosspoint density appreciably larger than those projected to be possible by the current evolution of the IC technology, but it is also consistent with the exploitation of functional molecules as memory elements. The ideal arrangement for the exploitation of bistable molecules is indeed based on their insertion in a crossbar structure.

4.1 The Crossbar Process

The current development of molecular electronics has most likely been possible only thanks to the fact that thiol-terminated molecules self-assemble spontaneously at the surface of noble or nearly noble metals (like gold, platinum, mercury, etc.) forming compact and ordered monolayers [98]. That self-assembled monolayers (SAMs) on preformed contacts of the said metals may behave as nanoscale memory elements was demonstrated for thiol-terminated π-conjugated molecules containing amino or nitro groups [99]. It does not surprise, therefore, that the first demonstration of nonvolatile molecular memories was given employing the crossbar structure functionalized with SAMs of thiol-terminated rotaxanes as reprogrammable cells [100]. The molecules were embedded between the metal layers forming the crossbar via a process, referred to as XB, that can be summarized as follows:

XB^1, deposition and definition of the first-level ("bottom") wire array
XB^2, deposition of a SAM of active reconfigurable molecules, functioning also as vertical spacer separating lower and upper arrays
XB^3, deposition and definition of the second-level ("top") wire array

Figure 4.1 sketches the XB process. The figure shows that whereas the crossbar structure is prepared via a planar technology, the memory cell, connecting top and bottom wires, is arranged vertically. The considerations of Sect. 3.3.3 show that the separation between upper and lower array must be higher than about 2.4 nm; if one wants that the information is not lost via quantum diffusion from one cell to a neighboring one, even the intercell separation must be larger than this limit.

Although potentially revolutionary, the XB approach (with double metal strips) has been found to have serious limits:

- The active organic element is incompatible with high-temperature processing, so that the top layer must be deposited at room or slightly higher temperature. This need implies a preparation based on physical vapor deposition (PVD), where the metallic electrode results from the condensation of metal *atoms* on the outer surface of the deposited organic films. This process, however, poses severe problems of compatibility, because isolated metal atoms, quite irrespective of their chemical nature, are mobile and decorate the molecule, rather than being held at its outer extremity [101–103].

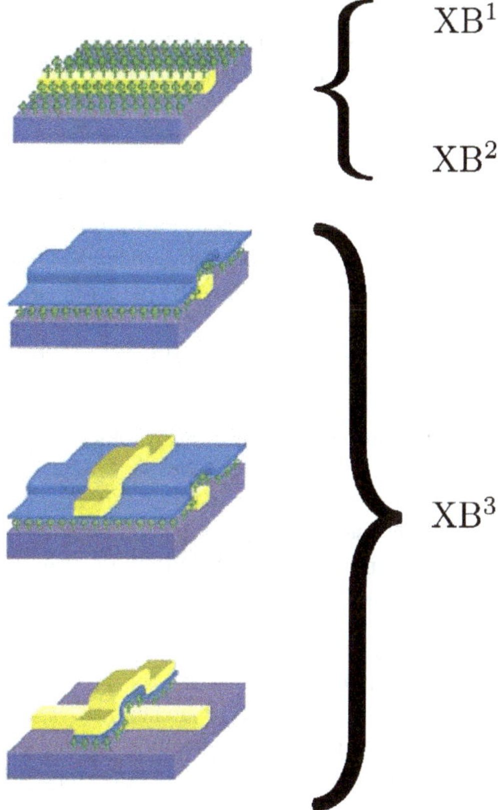

Fig. 4.1 The first proposed crossbar-architecture fabrication steps

- A safe determination of the conductance state of bistable molecules requires the application of a voltage V appreciably larger than $k_B T/e$. Since $k_B T/e = 25\,$mV at room temperature, a realistic lower limit to V is 0.1–0.2 V. Applied to molecules with typical length around 3 nm, this potential generates an electric field, of the order of $5 \times 10^5\,$V cm^{-1}, sufficiently high to produce metal electromigration along the molecules [104].
- The energy barrier for metal-to-molecule electron transfer is controlled by the polarity of the contact, in turn increasing with the electronegativity difference along the bond linking metal and molecule [105]. The use of thiol terminations for the molecule, as implicit in the XB approach, is expected to be responsible for high energy barriers because of the relatively high electronegativity of sulfur.
- The practical integration of noble or nearly noble metals with the MOS technology is very difficult [106].

The first difficulty can in principle be removed by slight sophistication of the process; for instance, one could rearrange the process specializing XB^3 as follows:

XB$^{3\,i}$, Spin coating the organic monolayer with a dispersion of metal nanoparticles in a volatile solvent

$XB^{3\,ii}$, evaporating the solvent, thus forming a relatively compact layer via coalescence of the metal particles

$XB^{3\,iii}$, compacting the resulting film by means of an additional amount of PVD metal

However, the other difficulties are more fundamental in nature and their solution requires the rejection of the SAM shibboleth. Although grafting functional molecules to silicon is a notoriously difficult job and the material properties of π-conjugated materials are still difficult to control, a solution to the electromigration problem can indeed be achieved by preparing the bottom electrodes in the form of silicon wires (as done in [107]), and the top electrodes in the form conducting π-conjugated polymers (as suggested in [108]). This new process, XB^1_+, can thus be summarized as follows:

XB^1_+, deposition and definition of the bottom array of polycrystalline silicon (poly-Si) wires

XB^2_+, deposition of the active (reconfigurable) element, working also as vertical spacer separating lower and upper arrays

XB^3_+, deposition and definition of the top array of conducting π-conjugated polymers

Process XB_+ is quite similar to that sketched in Fig. 4.1.

The use of poly-Si as material for the top array too seems impossible because it is prepared almost uniquely via chemical vapor deposition (CVD) at incompatible temperatures with organic molecules. The only way to overcome this difficulty consists thus in a process, XB^*, where the two poly-Si wire arrays defining the crossbar are prepared *before* the insertion of the organic element. Preserving a constant separation on the nanometer length scale is possible only via the growth of a thin film (working as a sacrificial spacer) on the first array before the deposition of the second [109–111]:

XB^1_*, preparation of a bottom array of poly-Si wires

XB^2_*, deposition of a sacrificial layer as vertical spacer separating lower and upper arrays

XB^3_*, preparation of a top array of poly-Si wires crossing the first-floor array

XB^4_*, selective chemical etching of the spacer

XB^5_*, insertion of the reprogrammable molecules in a way to link upper and lower wires in each crosspoint

The basic idea of process XB^*, based on the insertion of the functional molecules after the preparation of the crossbar, is sketched in Fig. 4.2.

Figure 4.3 shows a pictorial three-dimensional view of the structure as resulting after steps XB^3_* and XB^5_*.

Of the three considered processes (XB, XB^+, and XB^*), the one based on double-silicon wires is certainly the most conservative one and is thus expected to be easiest to integrate in IC processing.

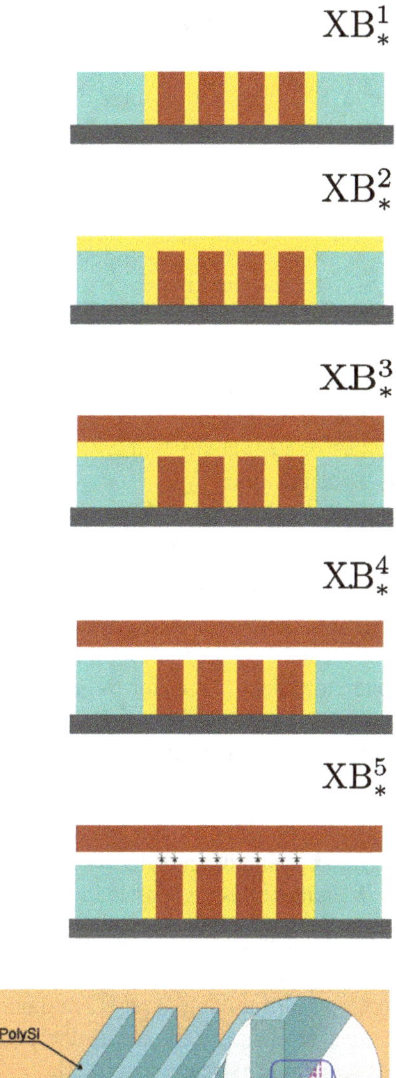

Fig. 4.2 The basic idea of XB$_*$: preparing the crossbar before its functionalization

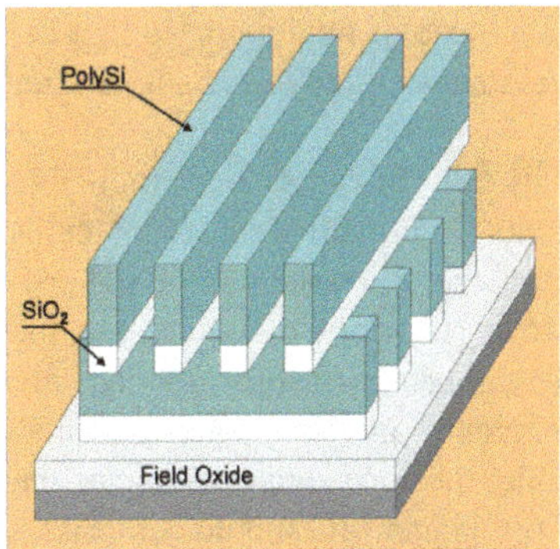

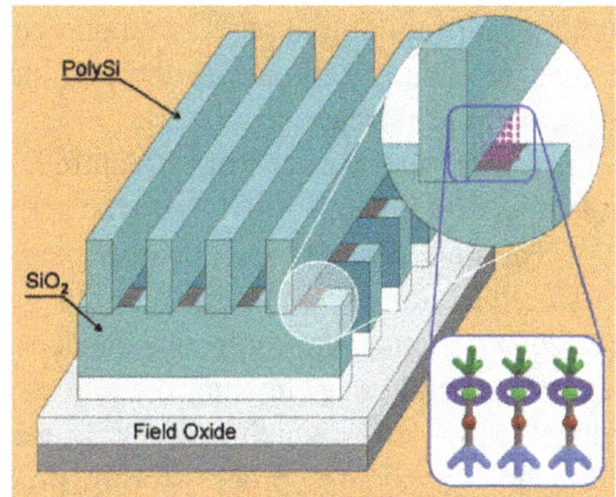

Fig. 4.3 Pictorial three-dimensional view of the structures resulting after preparation of a top array of poly-Si wires crossing the first-floor array (XB$_*^3$, *left*) and insertion of the reprogrammable molecules in a way to link upper and lower wires in each crosspoint (XB$_*$5, *right*)

4.2 Process Integration

Any CMOS process requires the combination of several unit operations. A nonexhaustive list is given in the following:

- Lithographic geometry definition
- Layer deposition of

  - Dielectrics (Si_3N_4 and SiO_2)
  - Metals (aluminum)
  - Semiconductors (poly-Si)
  - Polymers (resist)

- Wet or dry etching
- Ion implantation,
- Dry or steam oxidation
- Predeposition
- Diffusion
- Annealing in inert or reducing atmosphere

The last four unit operations are thermal ones, with largely variable temperature and duration, typically ranging from 450°C for about 30 min in a hydrogen atmosphere (for the annealing of Al:Si alloy) to about 1,000°C for several hours in an inert or oxidizing atmosphere (for p- or n-well diffusion). As sketched in Fig. 2.7, the total number of fabrication steps in a real process is between 300 and 500, a relevant part of which involves heat treatments. Any CMOS process can be thought of as chronologically organized in three parts, devoted to the formation of:

- Wells, insulations, and fields (at $T \gtrsim 950°C$)
- Sources and drains (at $T = 850–900°C$)
- Metallizations and passivations (at $T \simeq 450°C$)

The unit operations involved in the first two items are referred to as front-end; those in the third item as back-end. The unit operations are so organized that the heat treatments of the initial parts are much heavier than those of final parts, so that the dopant profiles resulting after the first steps remain substantially unchanged after subsequent heat treatments [112].

Assuming that the crossbar structure is formed by poly-Si wires (thus producible at temperature of 600–700°C) and the active elements are instead organic molecules (thus certainly destroyed at a temperature of 400°C), the following circuit and process architectures seem reasonable [113]. The circuit is formed by two zones: the *nanozone*, hosting the molecular devices, and the *lithozone* (in turn divided into active and field zone) hosting the silicon devices. Nano- and lithozones are interconnected by strips which become, in the nanozone, the rows and columns of the crossbar memory. The process for the production of the nano-IC should comprise a *zeroth mask* protecting (with Si_3N_4 or SiO_2) the nanozone from the *front-end initial stage*, where transistors, capacitors, and short-range poly-Si interconnections are lithographically defined. At a certain point in this stage of the process, the protection of the nanozone is removed and the poly-Si seed is allowed to cross the nanomicrofrontier. After protecting the active zone (again with Si_3N_4 or SiO_2) the *strip-formation stage* is started. After its conclusion, the nanozone is protected (once again with Si_3N_4 or SiO_2) and the *front-end final stage* is carried out. The process goes to completion as in standard IC processing. After that, the dielectric layer protecting the nanozone is selectively removed and the *grafting* of the active elements to the poly-Si wires is achieved by simple exposure to the organic molecules: the molecules are addressed to the wanted regions simply by the appropriate distribution of chemical potentials. In turn, that requires the ability to control the chemical terminations of the surfaces of the most important IC films. The passivation with organic resins of the overall nano-IC completes the process.

4.3 Why Molecules?

An analysis of the above situation shows that *most of the processing difficulties originate from the choice of organic molecules as active elements.* It is thus useful to understand which reasons motivate this choice.

Assume that the active element may be deposited as an inorganic film suitable as a memory element. Examples are ferroelectric materials in crossbar memories based on sensing capacitance, or materials exploiting the change of conduction resulting from phase change [114] or mimicking the memristor [115] in memories based on sensing current. In those cases the process, XB^{in}, could simply be constituted by the following sequence:

XB_1^{in}, deposition and definition of the bottom metal-wire array

XB_2^{in}, deposition of the active inorganic film, working also as vertical spacer separating lower and upper arrays

XB_3^{in}, deposition and definition of the top metal-wire array

In principle, not only is XB^{in} much simpler than XB, XB^+, and XB^*, but it is also more consistent with IC technology and easier to integrate. Thus, why should one consider molecules?

In the author's opinion, this choice is motivated by the scaling possibilities offered by the use of molecules, at least in the limit for which the conductance of a set of them in parallel is nothing but the sum of molecular conductances. This condition, however, may be violated for a lot of reasons; among them electrostatic effects seem the most important cause for deviation from linearity. Even ignoring interference phenomena for simultaneous electron flow along different molecules (most likely not occurring at room temperature because of decoherence), the change of conductance is generally associated with the generation or disappearance of charges in the molecule; each charge, while controlling the conduction along the hosting molecule, also influences the conductance of neighboring molecules because of the electrostatic field it produces.

If these effects can be ignored, the determination of the electrical behavior of the molecular ensemble on a certain length scale allows an almost rigorous extrapolation down to a single molecule. Thus, assuming one is able to produce a silicon–organic hybrid crossbar circuit on a given scale, *any improvement of definition limit, sense-amplifier sensitivity, etc., will allow the design and production of the circuit on a lower length scale.*

Assume, instead, that one has produced the crossbar where each active element is a nanoscopic system. Even assuming that one is able to produce an inorganic crossbar circuit on a given scale, if the system cannot be viewed as the union of weakly coupled parts (i.e., the case of ferroelectric, phase-change, or memristor materials), the $(N + 1)$ problem suggests that *any improvement of definition limit, sense-amplifier sensitivity, etc., will require a redesign of the circuit on a lower length scale.*

This discussion may be summarized concluding that *if scalability is a key point, the hybrid route is preferable.* Although scalability is an important issue, the practical realization of the hybrid route suffers from many difficulties. In particular, in addition to the problems listed by Keyes [11] and to the said reorganization of the process for IC production, the implementation of the silicon–organic hybrid architecture XB* requires indeed the solution of the following specific problems:

Problem 1: the setup of an economically sustainable technology for the preparation of crossbar with crosspoint density (say 10^{11} cm^{-2}) higher than the one projected by the Roadmap

Problem 2: the linkage of the addressing lines to the microelectronic circuit for writing and sensing

Problem 3: the design and synthesis of the reconfigurable molecules

Problem 4: the grafting of the functional molecules to those crosspoints via batch processing

The following four chapters are devoted to discuss how XB* can be adapted to solve all these problems.

Chapter 5
Crossbar Production

Problem 1 (Preparation of nanoscopic features) Preparing externally accessible arrangements of structures not producible by lithography, because of either technical reasons (below the lithographic limit) or economic reasons (excessive cost of advanced lithography).

As already discussed, the dramatic explosion of the cost required to sustain the reduction of the feature size (Fig. 2.6) has given rise to doubts on the possibility of continuing the current increase of density [by a factor of 2 each 18 months (24 months) for memories (microprocessors)] beyond the next decade. Hence the interest in the development of alternative lithographic methods or of nonlithographic patterning technologies for the definition of geometries on the nanometer length scale.

The crossbar structure has the potential to overcome the limits of conventional DUV, EUV, and EB lithographies because of the following reasons:

1. The fabrication of crosspoints does not require any alignment,[1] being the self-aligned result of two subsequent definitions.
2. The geometries of interest (usually parallel lines; see however Sect. 6.2) may be produced without any advanced lithography.

As for the first item, it was already observed (Sect. 2.1) that several of the major developments of the IC technology are marked by the invention of the following techniques for the *self-alignment* (rather than lithographic alignment) of a region on another:

- The *local oxidation of silicon* (LOCOS) technique [26, 27], for the self-alignment of active zones (where the transistors are built) on the field (the zones separating and insulating different transistors)
- The *silicon-gate technology* [25], for the self-alignment of the gate electrode on source-and-drain regions
- The *spacer patterning technology* (SPT) [29], for the self-alignment of the gate electrode with respect to source-and-drain metal contacts

---

[1] Of course, there remains the problem of aligning contacts to crossbar rows and columns; this problem will be the matter of Chap. 6.

Actually, the order indicated above is not the historical order in which they were discovered; rather, it is related to the way they are arranged in IC production processes.

Clearly enough, the crossbar structure can be prepared using standard or advanced (DUV, immersion DUV, EUV, or EB) lithographies. However, what is of special interest here is the fact that the extremely simple geometry involved in the crossbar, line arrays, allows its production without a massive use of EUV or EB lithographies. Arrays of parallel lines with the pitch on the nanometer length scale can indeed be prepared via nonconventional techniques; it is even more remarkable that the line widths and separations achievable with such techniques are smaller than the ones achievable via the most advanced lithographies [116].

These nonlithographic techniques (NLTs) exploit the following properties:

- The control of film thickness t (along, say, the "vertical" direction) is possible down to the subnanometer length scale, provided that the film is sufficiently homogeneous.
- Horizontal features (slabs) with thickness t can be transformed into vertical features (bars) whose width w is determined by t:

$$t \xrightarrow{\text{NLT}} w.$$

The preparation of sublithographic features with pitch on the 10-nm scale length goes back to the 1980s of the past century. The first attempts, however, were based on shadowing and lift-off techniques and of difficult practical exploitation [117, 118]. Techniques for the preparation of sublithographic features have attained a sufficient maturity only recently; among them, this book will deal with the *imprint lithography* (IL) and two variants of the *multispacer patterning technology* (S^nPT).

5.1 Imprint Lithography

Imprint lithography is a molding process in which the topography of a template defines the patterns over a substrate. This is achieved by means of a mechanical deformation of the resist materials used as hard mask layers. This renders the pattern replications completely free from the limiting factors of the lithographic equipments (such as light diffraction or beam scattering), thereby shifting the problem to the preparation of the mask and to the rheological control of the resist during pattern transfer. Actually IL is a *contact* (rather than proximity) lithography; what may be nonlithographic is the process used for the preparation of the mask (see [119] for an extended view of this matter). The major interest toward IL is as an alternative to standard photolithography and because of its potential of reducing the production cost of the masking process [119].

The first method for the nonlithographic preparation of nanoscopic line arrays for IL was proposed by Natelson et al. [120]. It is essentially based on the repeated

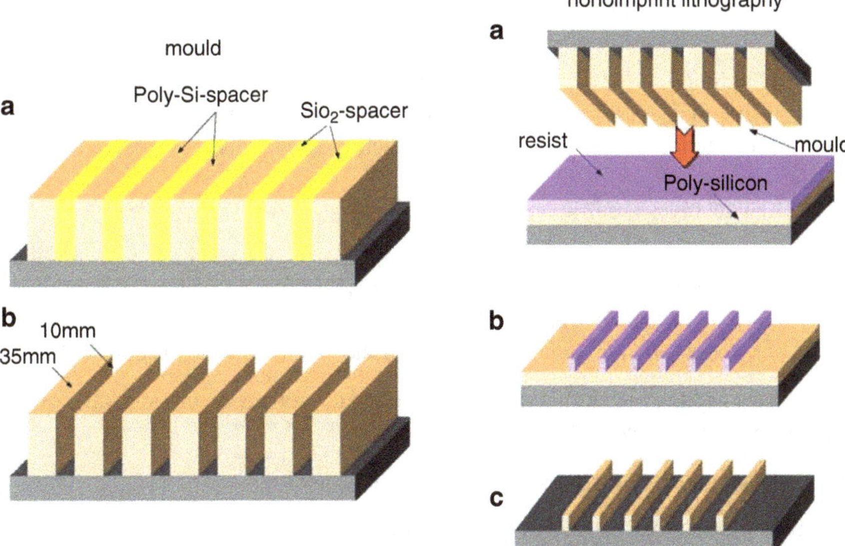

Fig. 5.1 Preparation of a mold via selective etching in multilayered structure (*left*) and its use as a contact mask for imprint lithography (*right*)

deposition of bilayer films, whose constituting materials, A and B, are characterized by the existence of a selective etching for one of them (say A). After cutting at 90°, polishing, and controlled etching of A, one eventually gets a mold that can be used as a mask for IL, as sketched in Fig. 5.1.

The first practical application of this idea to nanoelectronics was provided by Melosh et al., who prepared a contact mask for IL with pitch of 16 nm by means of the superlattice nanowire pattern transfer (SNAP) method. This method proceeds by growing a superlattice via molecular beam epitaxy, cutting the sample perpendicularly to the surface, polishing the newly exposed surface, and etching the different strata of the superlattice selectively [121].

A review of the SNAP method, not limited to the technology but extended to the functional characterization of silicon nanowires produced by this method, is given in [122].

Although IL is characterized by fast growth, little is known about the overall yield (preparation of mask and stamp, imprint, etching) of this process when the geometries are on the nanometer length scale. On the contrary, the other method considered in this chapter, the multispacer patterning technology (S^nPT), is based on a mature technology – the spacer patterning technology (SPT). This technology (as well as its multispacer variants) allows the nanowires to be patterned almost at will by growing them on a seed defined by a suitable mask. As discussed in Sect. 6.2, this property can be exploited for the demultiplexing.

5.2 Spacer Patterning Technology

The S^nPT admits two major variants; both of them are based on the spacer patterning technology. The SPT is an age-old technology originally developed for the dielectric insulation of metal electrodes contacting source and drain from the gate of MOS-FET.

The SPT involves the following steps[2]:

SPT^0, the *lithographic definition* of a seed with sharp edge and high aspect ratio

SPT^1, the *conformal deposition* of a conformal film of thickness as uniform as possible on this feature

SPT^2, the *directional etching* of the film until the original surface is exposed

If the process is stopped at this stage, it results in the formation of side walls of the original seed; otherwise, if

SPT^3, the original seed is removed via a *selective etching*.

what remains is only the walls of the seed edges.

Figure 5.2 sketches the various stages of SPT, without considering the actual shape of the conformally deposited film in the vicinity of the edge of the seed. The figure also shows the cross section of a poly-Si wire resulting after an SPT^0–SPT^3 cycle.

The SPT has been demonstrated to be able for the production of features with minimum size of 7 nm [123, 124] and has already reached a high level of maturity, succeeding in the definition of nanoscopic bars with high aspect ratio with yield very close to unity.

The SPT may be sophisticated due to the deposition of a multilayered film; Fig. 5.3 shows the sidewalls resulting from the deposition of a multilayer and compares it with what is really done in the original application of SPT – the insulation of the source-and-drain electrodes from the gate. In this case care was taken to account for the conformal deposition on the edge of the seed.

5.3 Multispacer Patterning Technology

The sharpness of the sidewall is controlled by the sharpness of the seed edge, the conformality of the deposition, and the directionality of the etching. If the sidewalls resulting after the completion of an SPT cycle are sufficiently sharp, the SPT can be reiterated. The multispacer patterning technology is just based on the repeated application of the SPT. Two S^nPT routes have been considered: the multiplicative ($S^nPT_\times$) and additive (S^nPT_+) routes.

---

[2] The reader is assumed to be familiar with concepts like "conformal deposition," "selective etching," "directional etching," etc.; however, their rigorous definitions is given in Chap. 13.

Fig. 5.2 (*Top*) The spacer patterning technology: SPT^0, definition of a pattern with sharp edges; SPT^1, conformal deposition of a uniform film; SPT^2, directional etching of the deposited film up to the appearance of the original seed; and SPT^3, selective etching of the original feature. (*Bottom*) Cross section of a poly-Si wire resulting from the application of the SPT

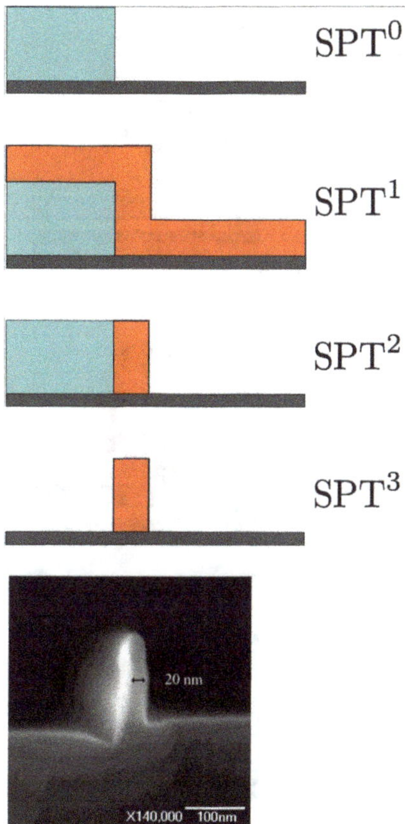

The $S^n PT_\times$ goes back to the 1980s: The first demonstrators were developed for the generation of gratings with sublithographic period [125]; its use for the preparation of wire arrays (in biochips) is in fact much more recent [124]. The $S^n PT_+$ is even more recent and was proposed having in mind the preparation of crossbars for molecular electronics [109–111].

The following part is devoted to a comparison of the limits and relative advantages of these routes on the basis of fundamental considerations.

5.3.1 Multiplicative Route: $S^n PT_\times$

The SPT allows the preparation of *two* spacers per seed [123]; in principle, this fact allows a multiplicative growth technology – $S^n PT_\times$. The multiplicative generation requires that both sides of each newly grown spacer are used for the subsequent deposition – that is possible only if the original seed is etched away at the end of any

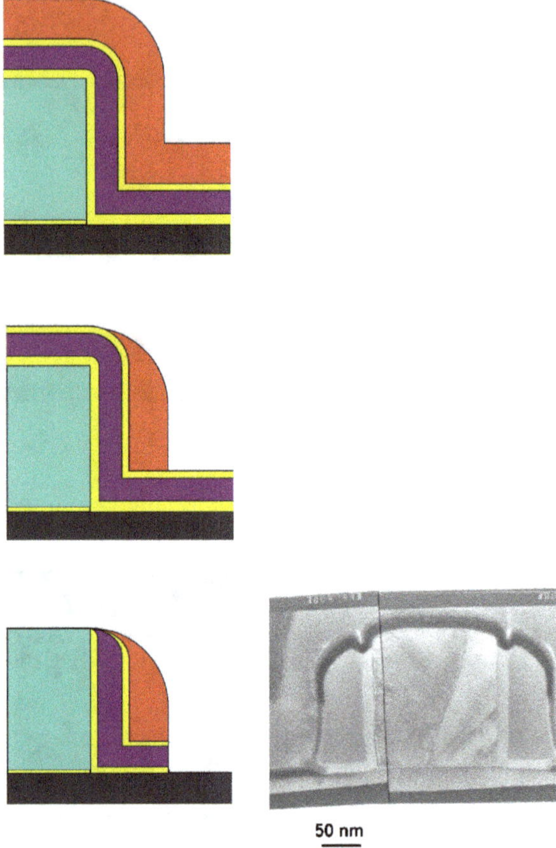

Fig. 5.3 (*Left*) The original application of the SPT in microelectronics – dielectric insulation of the gate from source-and-drain electrodes. (*Right*) Cross section of the resulting structure

cycle. Therefore, each $S^n PT_\times$ multiplicative cycle, $SPT_\times$, involves the following steps (as illustrated in Fig. 5.2):

$S^n PT_\times^1$, conformal deposition of a film on the seed

$S^n PT_\times^2$, directional etching of the newly deposited film up to the exposure of the seed

$S^n PT_\times^3$, selective etching of the original seed

Figure 5.4 sketches the idealized cross sections resulting during two $SPT_\times$ repetitions.

Assume that the process starts from a seed formed by an array with pitch P of lithographically defined seeds (lines) each of width W, the linear density K_0 of lines being thus given by $K_0 = P^{-1}$. The minimum value of P is determined by the considered lithography, while W is adjusted to the wanted value controlling exposure, etching, etc. Hereinafter, capital letters W and P and the corresponding

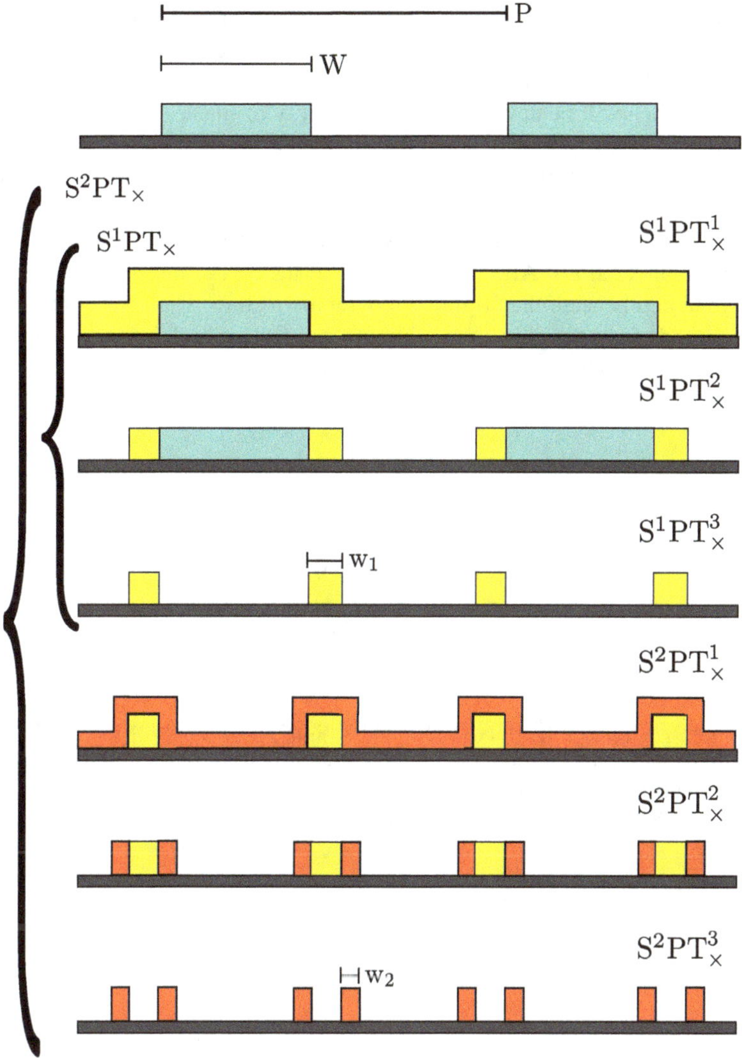

Fig. 5.4 Two $STP_\times$ cycles for the formation of a four-times denser sublithographic wire array starting from a lithographic seed array

lower cases, w and p, will be used to denote the same horizontal features on the litho- and nanoscale, respectively.

Let the seeds be formed by a given material A (to be concrete, think of it as SiO_2) and the $SPT_\times$ be carried out depositing another material B that can be etched selectively with respect to A (again for concreteness, B can be thought of as formed by poly-Si) thus defining a conformal layer of thickness $t_1 = \varrho_1 W$, with $\varrho_1 < 1$. After completion of the multiplicative SPT cycle, the surface will thus be covered by an array of $2K_0$ wires per unit length each of width $w_1 = t_1 = \varrho_1 W$.

Let the process proceed with the deposition of a film of A with thickness $t_2 = \varrho_2 w_1 = \varrho_1 \varrho_2 W$, with $\varrho_2 < 1$ (in the considered example, this process could be the oxidation of poly-Si to an SiO_2 thickness t_2). After completion of the second $SPT_\times$ cycle, the surface will be covered by a spacer array of linear density $2^2 K_0$ each of width $w_2 = t_2 = \varrho_1 \varrho_2 W$. It is noted that in $S^n PT_\times$ the seed material at the end of each $SPT_\times$ is inverted from A to B or vice versa, so that the material of the original seed could be chosen in relation to the parity (even or odd) of n.

The mask geometry that maximizes the spacer linear density is determined by the following considerations. After n reiterations of the $SPT_\times$, the spacers will extend both beyond and beneath the original lithographic feature. The zone containing the spacers extends from the edge of the lithographic feature both into the region separating them and into the region beneath the feature by amounts l^n_{out} and l^n_{in} given by

$$
\begin{aligned}
l^n_{\text{out}} &= w_1 + w_2 + \cdots + w_n \\
&= W \sum_{k=1}^{n} \prod_{j=1}^{k} \varrho_j,
\end{aligned}
\tag{5.1}
$$

$$
\begin{aligned}
l^n_{\text{in}} &= w_2 + \cdots + w_n \\
&= W \sum_{k=2}^{n} \prod_{j=1}^{k} \varrho_j.
\end{aligned}
\tag{5.2}
$$

The estimate of l^n_{out} and l^n_{in} requires the knowledge of the various ϱ_k. Without pretending to describe the actual technology, but simply to have quantitative (although presumably correct at the order of magnitude) estimates, ϱ_k is assumed to be independent of k, $\forall k$ ($\varrho_k = \varrho$). With this assumption (5.1) and (5.2) become

$$
\begin{aligned}
l^n_{\text{out}} &= W \sum_{k=1}^{n} \varrho^k \\
&= W \frac{\varrho}{1-\varrho} \left(1 - \varrho^n\right),
\end{aligned}
\tag{5.3}
$$

$$
\begin{aligned}
l^n_{\text{in}} &= W \sum_{k=2}^{n} \varrho^k \\
&= W \frac{\varrho^2}{1-\varrho} \left(1 - \varrho^{n-1}\right).
\end{aligned}
\tag{5.4}
$$

The least upper bounds of l_{out} and l_{in} are obtained by taking the limit for $n \to +\infty$ in (5.3) and (5.4): $l_{\text{out}} = \varrho/(1-\varrho)$ and $l_{\text{in}} = \varrho^2/(1-\varrho)$; moreover, for relatively low values of n, both ϱ^n and ϱ^{n-1} are already negligible with respect to 1 so that one can reasonably assume

$$
l^n_{\text{out}} \simeq W\varrho/(1-\varrho),
\tag{5.5}
$$

$$
l^n_{\text{in}} \simeq W\varrho^2/(1-\varrho).
\tag{5.6}
$$

For any W, the optimum ϱ is obtained by imposing the condition that all the region beneath the original lithographic feature is filled with nonoverlapping spacers: $2l_{\text{in}} = W$. Inserting this condition into (5.6) gives

$$\varrho \simeq \tfrac{1}{2}. \tag{5.7}$$

Similarly, the optimum size of the outer region is given by the following condition: $2l_{\text{out}} = P - W$. Inserting this condition into (5.5) gives

$$P \simeq 3W. \tag{5.8}$$

Choi et al. [124] has demonstrated that three $SPT_\times$ repetitions on a lithographically defined seed result in nanowire arrays of device quality and suggests that very long wires can indeed be produced with a high yield. However, even accepting that the process yield is so high as to allow the preparation of noninterrupted wires over a length (on the centimeter length scale) comparable with the chip size, if the lines are used as conductive wires of the crossbar, its length is so high as to have a series resistance larger than the resistance of the molecules forming the memory cell. This problem will be considered in the next chapter; it is however anticipated that its solution consists in the organization of the crossbar memory in modules each hosting a submemory of size 1–4 kbits. This implies that each module must be framed in a region sufficiently large to allow the addressing of the memory cells. In this way, the density calculated above is an upper value to the exploitable density.

5.3.2 Additive Route: $S^n PT_+$

The $S^n PT_+$ is substantially based on n STP repetitions where *the original seed is not removed and each newly grown bar defines a new edge of the seed for the subsequent STP*. Each SPT_+ cycle starts from an assigned seed and proceeds with the following steps:

$S^n PT_+^1$, conformal deposition of a conductive material
$S^n PT_+^2$, directional etching of this material up to the exposure of the original seed
$S^n PT_+^3$, conformal deposition of an insulating material
$S^n PT_+^4$, directional etching of this material up to the exposure of the original seed

The basic idea of the $S^n PT_+$ is shown in Fig. 5.5 [109–111]: the upper part sketches the process; the lower part shows instead how poly-Si wire arrays separated by SiO_2 dielectrics with sublithographic pitch (35 nm) can indeed be produced.

Consider an array with pitch P of lithographic seeds, each with width W (and thus separated from one another by a distance $P - W$). Denote with the same symbols in lower case, p and w, the corresponding sublithographic quantities. Starting from an array of lithographic seeds, after n repetitions of SPT_+, any seed is

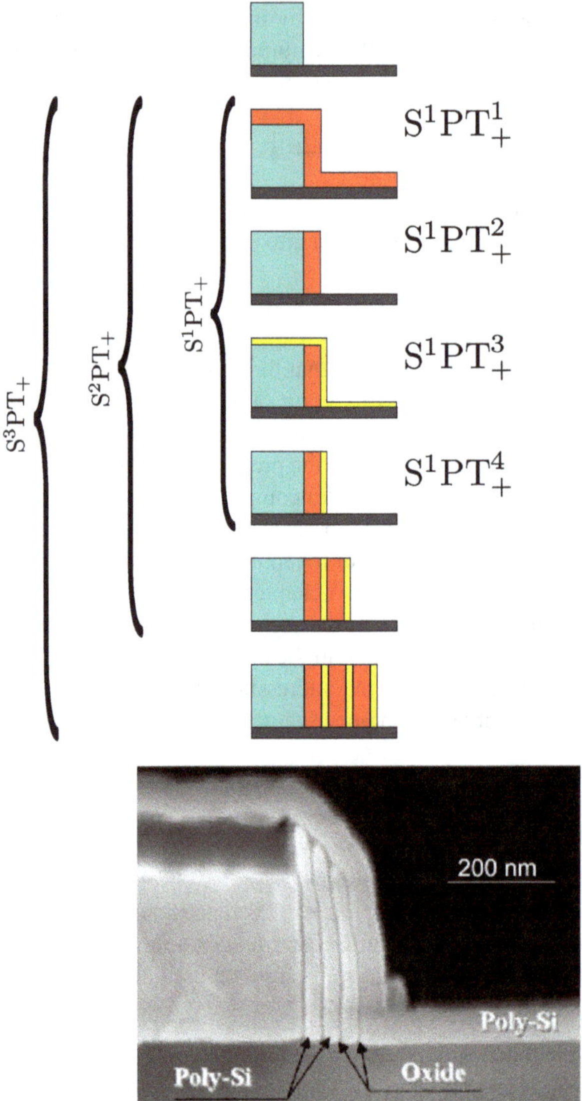

Fig. 5.5 *Up*: the additive multispacer patterning technology, split in its elementary steps ($S^1PT^1_+$– $S^1PT^4_+$) and after the second (S^2PT_+) and third (S^3PT_+) cycles. *Down*: an example of S^3PT_+ multispacer (with pitch of 35 nm and formed by a double layer poly-Si|SiO$_2$) resulting after three repetitions of the SPT$_+$ (taken from G.F. Cerofolini et al., *Nanotechnology* **16**, 1040 (2005))

surrounded by $2n$ lines (an example with $n = 4$ is shown in Fig. 5.6), so that the corresponding effective linear density K_n of spacer bars is given by

$$K_n = 2n/P.$$

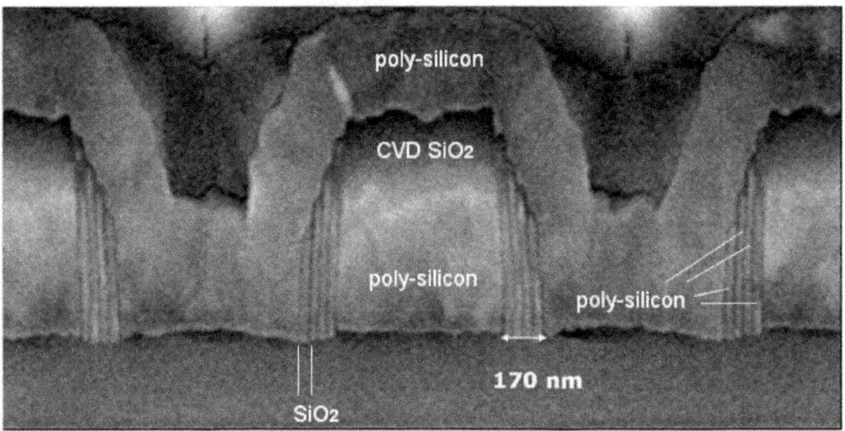

Fig. 5.6 An example of S^4PT showing, the construction of four silicon wires per side of the seed

The sketch in Fig. 5.5 shows a process in which lines are additively generated onto a progressively growing seed, preserving the original lithographic feature along the repetitions of the unit process. The unit process is based on two depositions of conformal layers (poly-Si and SiO_2) each followed by a directional etching.

However, Fig. 5.7 shows that a similar structure could be obtained by a cycle formed by

$S^n PT'_+$, conformal deposition of a two-layer film (formed by an insulating layer deposited before the conductive one – the order of deposition is fundamental)

$S^n PT''_+$, directional etching of this film up to the exposure of the original seed

The example of Fig. 5.6 shows also that the process can be tuned to preserve the constancy with n of bar width w_n, $w_n = w$, and pitch p_n, $p_n = p$. On the contrary, the bar heights s_n decrease almost linearly with n,

$$s_n \simeq s_0 - \tau n \tag{5.9}$$

(with τ the spacer height loss per SPT cycle), at least for n lower than a characteristic value $n_{\max}$.

This decrease is ultimately and unavoidably due to the fact that the conformal coverage of a feature with high aspect ratio results necessarily in a rounding off of the edge shape with curvature radius equal to the film thickness and that the subsequent directional etching produces a nonplanar surface. Figure 5.8 explains the reasons for the shape of the resulting bar.

Even assuming a perfectly directional etching, the resulting spacer is not flat and there is a loss of height τ not smaller than t: $\tau \geq t$. Figure 5.5 shows that the process can actually be controlled to have τ coinciding, within error, with its minimum theoretical value, $\tau = 1.0\,t$. However, Fig. 5.6 shows that τ depends on

Fig. 5.7 The additive multispacer patterning technology can be modified to reduce the number of directional etchings by a factor of 2 via sequential deposition of a bilayered film. There are two variants: deposition first of poly-Si and then of SiO_2 (*left*), and of the same layers in reverse order (*right*)

the process, and this can be tuned to have $\tau = 3.2\,t$. Although the loss of height may seem a disadvantage, in Sect. 6.3, it will be shown how a controlled decrease of s_n with n may be usefully exploited.

Assuming the validity of (5.9) until s_n vanishes, the maximum number $n_{\max}$ of SPT_+ repetitions is given by $n_{\max} = s_0/\tau$; after $n_{\max}$ SPT_+ repetitions, the seed is lost and the process cannot continue further. In view of the availability of techniques for the production of deep trenches with very high aspect ratios, in this analysis, s_0 (and hence $n_{\max}$) may be regarded as an almost free parameter.

The optimum distance allowing the complete filling of the void regions among the original lithographic seeds is therefore given by $P - W = 2n_{\max}p$. Hence, the maximum number of crosspoints that can be arranged in any square of side P is given by $(2n_{\max})^2$, and the maximum effective crosspoint density $\delta_{\max}$ achievable with the S^nPT_+ is given by

$$\delta_{\max} = (2n_{\max}/P)^2$$
$$= \frac{1}{p^2}\left(\frac{1}{1 + W/2n_{\max}p}\right)^2. \tag{5.10}$$

Fig. 5.8 Ideal shape of a sidewall after conformal deposition (*top*) and directional etching (*bottom*)

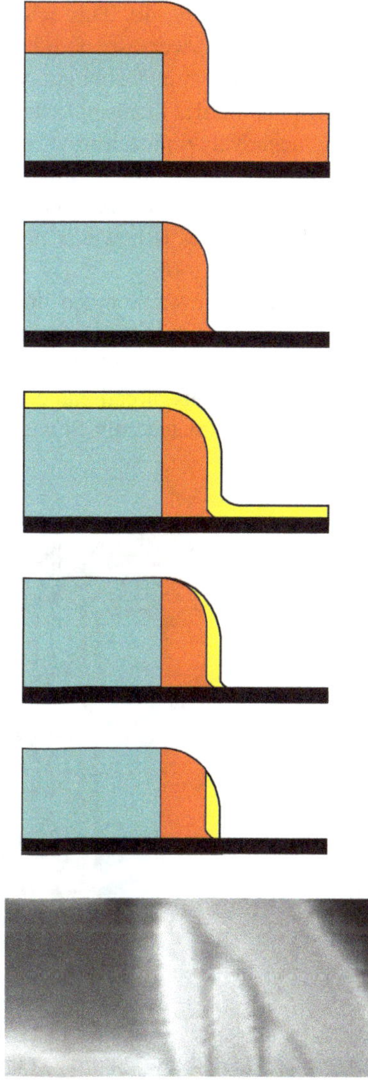

Equation (5.10) shows that δ_{max} depends on the lithography (through W) and on the sublithographic technology $S^n PT_+$ (through n_{max} and p).

The rate τ of height decrease depends on how accurately the technology has been set; just to give an idea of the maximum achievable density, Fig. 5.5 shows that $p = 35\,\text{nm}$ has already been achieved and $n_{max} \simeq 10$ is at the reach of the $S^n PT_+$; for $W = 0.1\,\mu\text{m}$ (characteristic value for high-volume IC production) and $p = 30\,\text{nm}$, (5.10) gives $\delta_{max} \simeq 8 \times 10^{10}\,\text{cm}^{-2}$. The comparison of this prediction with the lithographically achievable crosspoint density (currently of about $2 \times 10^9\,\text{cm}^{-2}$) shows that $S^n PT_+$ allows the crosspoint density to be magnified by a factor of about 40. This is however achieved only with the construction of ten consecutive spacers per both bottom and top layers. Since the spacer patterning technology is a mature technique (with yields close to unity when employed in more complex geometries than single lines), its repeated application presumably does not impact negatively on feasibility. However, so many repetitions influence processing cost; the increase of cost implied by its repeated application in IC processing has been discussed in [126]. Although this increase is moderate, the integration of so many SPT_+ cycles is however not trivial and passes through the development of dedicated cluster tools.

It should however be noted that the basic idea sketched in Fig. 5.7 can be extended to remove this difficulty at least partially: the conformal deposition of a single slab formed by n poly-Si|insulator bilayers (the insulator being SiO_2, Al_2O_3, etc.) followed by its directional etching would indeed result in the formation of $2n$ dielectrically insulated poly-Si wires (see Fig. 5.9).

Fig. 5.9 Structure resulting after the deposition of a slab of four poly-Si|SiO$_2$ bilayers (*top*) followed by the conformal attack stopped with the exposure of the original lithographic seed (*bottom*)

The figure however shows that this dramatic simplification of the process can be done for the deposition of the bottom array only.

5.4 Minimum Exploitable Bar Width

The examples mentioned previously show that IL and S^nPT_+ succeed in the preparation of arrays with pitches of 16 and 30 nm, respectively. It is also likely that the use of subtractive techniques or atomic layer deposition will allow S^nPT_+ to achieve the same limits as IL. In principle, techniques like molecular beam epitaxy, atomic layer deposition, and oxidation permit further reduction of the above quantities. However, could it be effectively applied in molecular electronics?

A technological answer to this question was already given in [127]: the detection of the conduction state of the molecules in each crosspoint requires the measurement of the current flowing through it under a certain applied voltage. In the hybrid approach this is done by an integrated amplifier. If this amplifier can sense a current $I_{\min}$, the minimum crosspoint area $A_{\min}^{XP}$ is given by

$$A_{\min}^{XP} = I_{\min}/J_{\max}^{ON},$$

where $J_{\max}^{ON}$ is the maximum current density flowing through the crosspoint. At the forefront of technology $I_{\min} \simeq 1\,\mu A$ whereas the most conducting reprogrammable molecules have $J_{\max}^{ON} \simeq 1\,\text{nA/nm}^2$, so that one gets $A_{\min}^{XP} \simeq 10^3\,\text{nm}^2$, corresponding to a minimum bar width $w_{\min}$ of 30 nm. Since the given examples of S^nPT_+ have already achieved this limit and IL is well below it, the above result would suggest that at this moment, reducing the pitch further is a useless exercise of stressing technology.

However $I_{\min}$ is strongly dependent on the current technology and on sensing strategy while $J_{\max}^{ON}$ is higher than its theoretical limit by so many orders of magnitude (five orders, as discussed in Sect. 7.1.1) that progress is expected. The combination of these factors suggests that minor deviations from the original assumption may result in an appreciable lower value of the minimum width $w_{\min}$.

For metal wires, a more physical limit is obtained, observing that when the bar width is reduced to a value for which quantum size effects are manifested, the unavoidable local width fluctuation ("horizontal roughness") will produce dramatic changes in bar conductance.

For semiconducting wires, one has to consider the bandgap widening in restricted geometries. This effect is expected to be manifested in values of w for which quantum size effect is still negligible. The amount of this effect may be estimated by observing the confinement of a dopant in a region. It is responsible for an increase of its ionization energy from the value E_∞^{ion}, which is characteristic of the extended solid, to a value E_w^{ion}. It is determined by the fact that the orbital of the neutral dopant is eventually limited by the size w of the confining region. Describing the dopant as a hydrogen-like atom, one has

$$E^{\mathrm{ion}} \simeq E^{\mathrm{ion}}_{\infty} \left(1 + \frac{a_{\infty}}{w/2}\right)^2$$
$$= E^{\mathrm{ion}}_{\infty} \left(1 + 2a_{\infty}/w\right)^2,$$

where a_{∞} is the hydrogen-like Bohr radius of the dopant in an unrestricted geometry. If this relationship does really hold true, the dopant is expected to be effectively frozen when w approaches $2a_{\infty}$. Since the conductance of semiconductors can be controlled by doping only when dopant ionization is possible, the previous considerations suggest $w^{\mathrm{min}} \simeq 2a_{\infty}$. For silicon $a_{\infty} \simeq 3\,\mathrm{nm}$, so that going below a size around 6 nm is expected to result in properties that are not scaled from those observed on a larger scale. On this length scale, however, the band properties of silicon are no longer the same as those of bulk silicon; evidence for this is obtained from photoluminescence properties of nanowires [128].

Chapter 6
The Litho-to-Nano link

Problem 2 (Litho-to-nano link) Linking in a one-to-one fashion features with separation below the lithographic limit to lithographically defined contacts.

If the availability of nanofabrication techniques is fundamental to establish nanotechnology, no less vital is the integration of the nanostructures with higher-level structures. As emphasized in a scholium to [129], "the difficulties in communication between the nanoworld and the macroworld represent a central issue in the development of nanotechnology." This is particularly true in the case of hybrid silicon-molecular devices, wherein once the crossbar structure is formed, it is necessary to link it to the embedding conventional CMOS circuitry [130].

The importance of addressing nanoscale elements in arrays goes beyond the area of memories and will be critical to the realization of other integrated nanosystems such as chemical or biological sensors, electrically driven nanophotonics, or even quantum computers.

In the following analysis, however, attention will be confined to the problem of addressing, by means of externally accessible lithographic contacts, nanoscopic crosspoints in a sublithographic crossbar. The practical exploitation of the functionalized crossbar requires that once this structure has been prepared, it must be linked to conventional silicon circuitry. This is a difficult task because the current conventional CMOS circuits at the forefront of the technology are characterized by a feature size of 45 nm (half pitch) whereas crossbars can now on be prepared with a pitch of, say, 30 nm.

External accessibility to the molecular world requires a hierarchy of devices and addressing protocols. The paradigmatic example is that of a crossbar memory, arranged in about N^2 crosspoints, defined by $2N$ nanoscopic wires, addressed by $\log_2(2N)$ lithographically defined gates. For $N = 10^5$, this gives 10^{10} crosspoints, 2×10^5 lines (the mesoscopic scale), and $1 + 5\log_2(10)$ addressing gates.

To evaluate the amount of conventional silicon circuitry required to operate the crossbar, consider an arrangement of 10^{10} bits, addressed by $10^5 + 10^5$ wires, and sensed by $2^8 = 256$ sense amplifiers. Assuming that each line is addressed with a gate formed by 4 transistors, the number of transistors required to manage the information stored in the crossbar is thus of the order of (though smaller than) 10^6. This would imply a crossbar area of the order of $0.25\ \mathrm{cm}^2$ for a density comparable to the arrangement sketched in Fig. 6.4 and a conventional CMOS area strongly depending on the memory architecture but likely smaller than $0.1\,\mathrm{cm}^2$ at the current CMOS density. Repeating the same argument for a memory of 10^{11} bits, one would obtain a crossbar area of $2.5\,\mathrm{cm}^2$ and a conventional CMOS area smaller than $0.3\,\mathrm{cm}^2$.

Several strategies have been adopted to attack the problem of the litho-to-nano link:

- In one of them, the complicated circuitry required by binary-code demultiplexing (see Sect. 6.4) is avoided using nanoscopic addressing circuits, embodying in themselves the ability to address single cells. In the first demonstrator, they were prepared by a "vertical" growth of variously doped silicon wires, first chemical vapor deposited and then arranged horizontally to link the nanoscopic lines to the lithographic lines [131–133]. The precise way such links cover the nanoscopic lines is not exactly known as this technique does not allow one to know a priori which cell is addressed. However, "this was a rudimentary demonstration vehicle, and neither vehicle nor probably the programming technique could be extended to the much larger arrays required for practical interest" [134]. A conceptually similar technique was however developed by the collaboration of [134], using a consistent solution with CMOS technology. In this solution, where a one-to-one correspondence between nano- and lithowires is achieved via rotation by a very small angle, the lithographic lines need to be defined with the same pitch as the nanoscopic lines.
- The idea of using small misalignment to access the nanoscopic circuitry was perhaps first proposed by Likharev with the name of CMOL circuit (see [135–137] for recent reviews). In Likharev approach the contact is achieved with the growth of pyramids whose base is matched to the lithographic feature and whose summit is matched to intersect one and only one nanoscopic line. The misalignment guarantees this occurrence, but makes it impossible to know the correspondence between crosspoint and wire pair.
- A misalignment-tolerant random-interconnect architecture was hypothesized by the proponent of the crossbar structure [93, 138]. In this way a large amount of cells are left out of control (being either not addressed or such that different cells are identified with a unique code) and a special logic must be employed to manage this situation, eliminating unaddressed cells and possibly exploiting redundancy (the Teramac approach[1]) [140]. Needless to say, this design is convenient only if the nonfunctioning cells are a minority and the circuitry required to manage the randomness remains relatively simple. That these conditions may really be satisfied in practice is not clear, especially if the expected crosspoint density ($10^{11}\,\mathrm{cm}^{-2}$) is compared with the soon achievable density ($10^{10}\,\mathrm{cm}^{-2}$) for NAND memories, whose control is deterministic.

---

[1] The Teramac was an experimental massively parallel computer designed in Hewlett–Packard laboratories in the 1990s. Contrary to traditional systems, which are useless if there is one defect, Teramac used defective processors – intentionally – to demonstrate its defect-tolerant architecture. Even though the computer had 220,000 hardware defects, it was able to perform some tasks 100 times faster than a single-processor high-end workstation [139].

In the following sections, attention will only be focused on arrangements allowing a deterministic access to all crosspoints and using two length scales – the sublithographic one (possibly down to a few nanometers) and the lithographic one. In particular, three methods will be discussed – one suitable for all kinds of nanoscopic crossbars (irrespective of their preparation procedure) and the other two specific for crossbars prepared by $S^n PT_+$.

6.1 The Horizontal Beveling Technique

The first technique is reminiscent of the beveling technique once used, together with microsectioning and staining, for the measurement of junction depth (and more recently of carrier concentration profiles using spreading resistance techniques); it will henceforth be referred to as horizontal beveling technique (HBT).

Consider as above an array of n conductive parallel wires aligned in the x direction with width w_y and pitch p_y, so densely arranged that they cannot be singularly accessed photolithographically and thus cannot be linked to the hosting microelectronic circuit.

A kind of hardware demultiplexing is possible via the following process: After the deposition of a protective insulating cap, a line with width b, oriented in a direction tilted by a small angle α with respect to the x direction and so long as to cross all wires, is defined and this pattern is used to etch the underlying insulating cap (see Fig. 6.1). This process will result in the exposition of n rhomboids with pitch P'_x and major side w'_x in the x direction given by

$$P'_x = p_y / \tan \alpha, \tag{6.1}$$

$$w'_x = b / \sin \alpha, \tag{6.2}$$

Fig. 6.1 The trick adopted for contacting each wire in the horizontal beveling technique

the area of each open contact being thus $bw_y/\sin\alpha$. The orientation is chosen in such a way that P_x' is in the reach of (not necessarily electron beam) lithography. Because of (6.1) and (6.2), this operation is certainly possible for sufficiently small α. Since the method produces from an array with pitch p_y and another array with pitch P_x' [controlled uniquely by the orientation angle α of the cutting bar, see (6.1)], it may viewed as a kind of *pitch converter*. This method is immediately extended from one to two arrays by defining two cutting lines (one per row and one per column).

Even this method, proposed by Cerofolini and Mascolo [126], is based on controlled misalignment. Compared with the one proposed by Likharev, it has the advantages of greater consistency with the current IC technology and of specifying exactly the location of the addressed crosspoint. Since each crosspoint can be addressed independently of the remaining ones, this architecture is suitable for random-access memories (RAMs).

6.2 Fusing Adjacent Lines in S^nPT_+

After the demonstration that the S^nPT_+ can be used for the preparation of nanowire arrays, Cerofolini and Mascolo proposed another addressing strategy exploiting specific features of this technology [126]: The process starts with deposition of a film (e.g., of Si_3N_4) and its photolithographic definition to form a seed with the pattern shown in Fig. 6.2a. The pattern is derived from the rectilinear edge required for the multispacer preparation with the addition of n rectangular indentations, of sides b_k ("width," parallel to the border) and a_k ("depth," perpendicular to the border); k runs from 1 to n and enumerates the wire in the array. The pitch B_k describing the indentation sequence is assumed to be given by

$$B_k = b^0 + b_k, \tag{6.3}$$

where b^0, the separation between vicinal indentations, does not vary with k; instead, a_k is assumed, at least temporarily, to be independent of k and given by

$$a_k = a^0 + (n-1)t_{sp}, \tag{6.4}$$

where t_{sp} is the width of the double spacer formed by a poly-Si spacer (with width t_{Si}) and an SiO_2 spacer (with width t_{ox}), $t_{sp} = t_{Si} + t_{ox}$, and the meaning of a^0 will be discussed later.

In S^nPT_+, alternate spacers of poly-Si and SiO_2 are grown after conformal deposition and directional etching. This process results in the reproduction indentations, only if $b_n > 2nt_{sp}$. If this condition is not satisfied, at a certain stage of the S^nPT_+ the indentation is completely filled. Suppose the indentation width b_k satisfies the following condition:

$$b_k = 2t_{Si} + 2(k-1)t_{sp}. \tag{6.5}$$

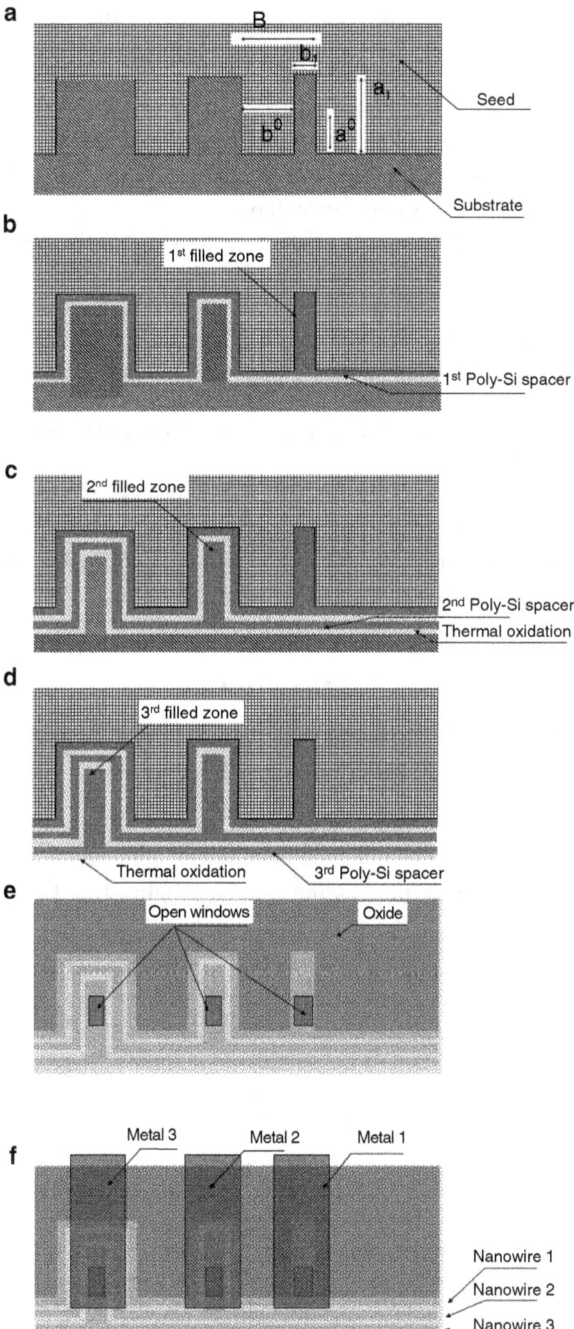

Fig. 6.2 Plan view of the mask that allows single wires to be contacted using the multispacer patterning technique

The first $S^n PT_+$ step fills the first indentation (Fig. 6.2b); the inspection of Fig. 6.2 shows that the occurrence of this phenomenon extends progressively to the kth indentation with the order k of bar deposition. After step k, the central region of each indentation up to the kth is filled with a poly-Si bar of width $2t_{Si}$; the remaining part of the cave is filled with $(k - 1)$ double layers each of width t_{sp}. At the end of the $S^n PT_+$ (Fig. 6.2d) all indentations have their central zone connected with only one nanowire running out of the original indentation.

After the $S^n PT_+$ completion, the process continues with the deposition of a dielectric film (SiO_2) on the whole structure and with the definition of n rectangular windows (of nominal sides a^0 and width $2t_{Si}$) aligned on the central zone of each cave (Fig. 6.2e). The process is concluded with the deposition of a metal film and the definition of n metal strips contacting the single nanowires.

The function a^0 and b^0 is now clear: a^0 controls the area metal–poly-Si contacts, whereas b^0 is determined by the minimum allowed distance between metal strips (Fig. 6.2f).

The process described above has the great advantage of allowing the definition of the litho-to-nanocontacts simultaneously with the formation of the wire array. It is however stressed that the considered architecture is quite demanding about the mask. To evaluate the tolerance it is noted that the maximum width b_k^{max} of the kth indentation must not be so large as to allow the formation therein of filaments of the $(k + 1)$th wire:

$$b_k^{max} = 2kt_{sp}; \tag{6.6}$$

the minimum width b_k^{min} of the kth indentation must instead be so large as to guarantee that the kth wire therein is certainly connected to the wire forming the crossbar:

$$b_k^{min} = 2(k - 1)t_{sp} + t_{Si}, \tag{6.7}$$

where the wire continuity has been assumed to be certain when the wire has a thickness t_{Si}. Combining (6.6) and (6.7) one has

$$b_k^{max} - b_k^{min} = 2t_{sp} - t_{Si}$$
$$= t_{sp} + t_{ox}; \tag{6.8}$$

In the case of Fig. 5.5 (with $t_{sp} = 35\,nm$ and $t_{ox} = 5\,nm$) (6.8) gives a tolerance, of $\pm 20\,nm$, compatible with current technology. It is however noted that the mask opening the contacts in the worst case (6.7) should have a maximum size (along the wire) of just $t_{sp} + t_{ox}$; in the example of Fig. 5.5, $t_{sp} + t_{ox} = 40\,nm$, which is inconsistent with the current technology. This value would run in the range achievable with standard photolitography taking a larger value of t_{ox}.

The described procedure is not optimized. Area occupation can be reduced taking a_k varying with k, instead of the constant value $a^0 + (n-1)t_{sp}$ in (6.4). For instance, choosing

$$a_k = a^0 + (k - 1)t_{sp} \tag{6.9}$$

still allows the litho-to-nano link thereby saving more space.

The crossing of this n nanowires with the analogue ones, perpendicularly oriented and lying on a parallel plane (obtained by the deposition or growth of an oxide of thickness adapted to the molecular size) results in the crossbar. Since the access to each single crosspoint is independent of others, the described addressing strategy is suitable for application to RAMs .

6.3 Energetic Filtering

Whereas in IL all wires are defined collectively, in $S^n PT_+$ they are constructed sequentially. The batch fabrication would make IL preferable over $S^n PT_+$ unless one were able to *use the sequential array deposition for the external recognition of single wires.*

An inspection of the right-hand side of Fig. 5.6 shows that $S^n PT_+$ results in wires whose height is progressively reduced with the order of preparation. Cerofolini proposed to exploit this spatial difference for the recognition and separation of single nanoscopic wires by electrical probes [127].

While the spatial and functional organizations of the crossbar are irrelevant for the methods described in the previous section, they are essential for the method described in the following. The entire crossbar is divided into M modules each formed by a subcrossbar of $n \times n$ wires (with $N^2 = Mn^2$), each module being obtained by filling of a (judiciously lithographically defined) trench via $S^{n/2}PT_+$ with $n/2$ lower than the value $n_{\max}$ considered in (5.10).

Needless to say, the problem of addressing the nanometer sized crossbar is reduced to that of addressing a single module, provided that it contains a sufficiently large (i.e., provided that $n^2 \gg 1$) number of nanoscopic crosspoints. Our attention will therefore be concentrated on addressing each module. Let r_i and c_j denote the ith row and jth column, respectively, and (r_i, c_j) be the crosspoint they define. All the rows r_i (columns c_j) run below a photolithographically defined electrode GE_i (GE_c) and finally terminate at a unique photolithographically defined contact C_r (C_c), as it is well represented in the cross sections in Fig. 6.3. The process is arranged in a way that electrodes GE_r and GE_c are separated from the underlying wires by an insulator of fixed thickness. Exploiting the electrodes as self-aligned masks, the nanowires are heavily doped everywhere but below the electrodes via ion implantation and subsequent thermal activation of a suitable dopant.

This configuration reminds one of a transistor in depletion mode, where GE_r and GE_c operate as control gates, the underlying slightly doped sector of nanowire as the semiconductor channel and the adjacent highly doped sectors as drain and source. If no voltage is applied on gates with respect to the body, the carriers are free to move in the whole nanowire; if a repulsive voltage is applied, the channel is depleted and there is no current flow between C_r and C_c. In principle, the voltage necessary to disable a nanowire decreases with its height.

For assigned poly-Si nanowire doping and insulator thickness, there exists at least one couple of voltage values of GE_r and GE_c which does not totally deplete the

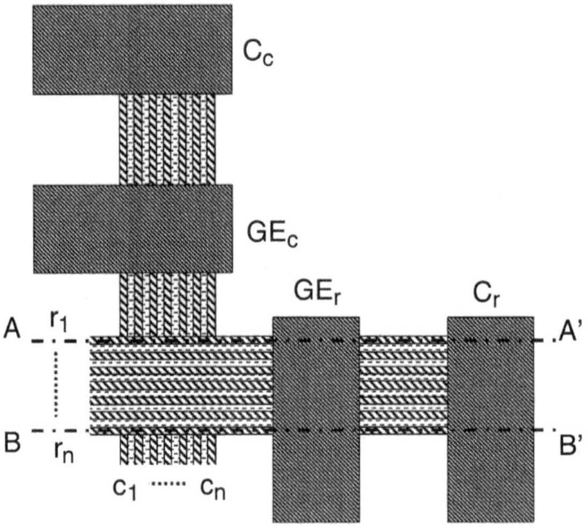

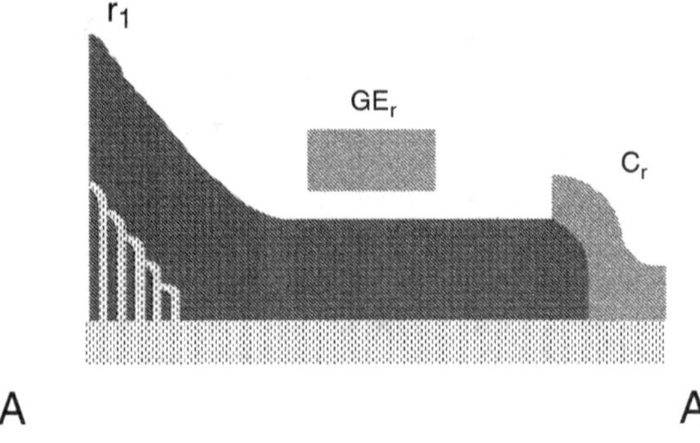

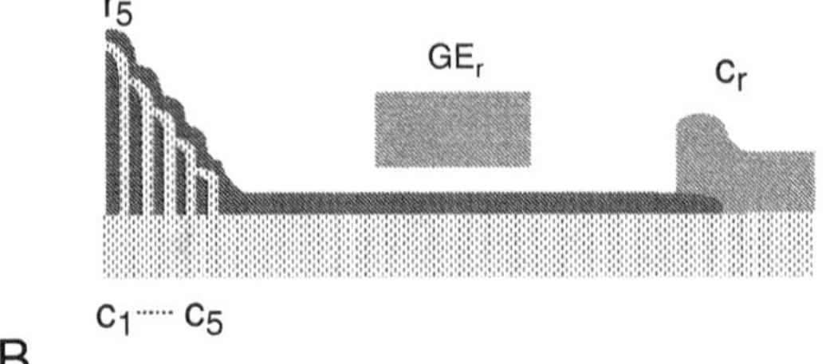

Fig. 6.3 Plan view of the crossbar and of the lithographically defined contacts (*up*) and views projected on planes AA' and BB' perpendicular to the surface and crossing the tallest and smallest row lines, respectively (*down*)

highest wires, r_1 and c_1. In this case, the current $I_{(1,1)}$ flowing from the contact C_r to the contact C_c is controlled by the conduction state of crosspoint (r_1, c_1) alone. If the voltage of GE_c is modified in such a way that even the wire immediately shorter, c_2, is enabled, the current flowing from C_r to C_c becomes the sum of the current $I_{(1,1)\cup(1,2)}$ flowing through (r_1, c_1) and (r_1, c_2). If $I_{(1,1)}$ was memorized, the current $I_{(1,2)}$ flowing through (r_1, c_2) (giving its state) is obtained as difference $I_{(1,1)\cup(1,2)} - I_{(1,1)}$.

Reiterating the argument to all crosspoints, the states of each of them can be determined, provided the information contents of all crosspoints are memorized in a buffer memory. This implies that only a small part of the crossbars can be read in parallel, because each one needs the availability of a circuitry for analog-to-digital conversion and for the manipulation of the information.

By the time, the addressing system depends only on the different height of the nanowires induced by the $S^n PT_+$. Some modification can be introduced in order to optimize the process. If the poly-Si wires are doped singularly (for instance, via gasphase doping during each poly-Si deposition), the difference in the threshold voltage from wire to wire may be magnified. Also, the oxide thickness between wires and electrodes GE_r and GE_c can be varied, thus resulting in another possibility of tuning the threshold voltage.

The addressing protocol would be greatly facilitated if the molecules operating as a memory element were a series of a reprogrammable moiety and of a resonant tunnel diode [141]. In that case, its operation as a Schmitt trigger would indeed allow the direct addressing of single bits without additional elaboration of data.

Remember now that both (lower and upper) nanoscopic arrays are more conveniently prepared for carrying out the $S^{n/2}PT_+$s in a cave, because the repetition of n SPT_+ produces n, rather than $n/2$ wires. This process, however, results in the formation of n pairs of wires with the same height, so that a single control electrode covering all wires would be unable to resolve those with the same height. This difficulty may however be circumvented by assigning to the cave a size, allowing the growth of just n wires separated by a lithographically accessible distance, and using two control-gate electrodes (denoted GE_l^u and GE_l^d, $l = r, c$) per line, each covering totally n electrodes.

Of course, the method described above works on the hypothesis that the resistors and FETs built on nanowires behave as expected by scaling their behavior on the length scale of 10^2 nm. Although physical limits do not seem to forbid scaling to a length scale of 10–30 nm (eventually leading to circuits with bit density of higher than 10^{11} cm^{-2}), tuning the technology for that might be a major problem.

6.4 Technology and Architecture

While the repetition in the additive way of n SPTs per (bottom and top) layers magnifies the lithographically achievable crosspoint density K_0^2 by a factor of $(2n)^2$, the repetition in multiplicative way gives a magnification of 2^{2n}. The magnification factor increases quadratically with the additive way and exponentially with the multiplicative way.

To estimate numerically the process simplifications offered by $S^nPT_\times$ over SPT_+, consider for instance the case of the three $SPT_\times$ repetitions per layer. This would produce a magnification of the lithographic crosspoint density by a factor of $2^3 \times 2^3$. Taking $W = 0.1\,\mu m$, after three $SPT_\times$ repetitions, the spacer width should be of 12.5 nm, with minimum separation of 25 nm. Taking into account (5.8), the crosspoint density achievable with the repetition of 2×3 $SPT_\times$ would thus be almost the same as that obtainable with the repetition of 2×10 SPT_+ (7×10^{10} vs. 8×10^{10} cm^{-2}).

Figure 6.4 shows in plan view a comparison between the following crossbars:

(a) A 2×2 crossbar obtained by crossing lithographically defined lines
(b) A 16×16 crossbar obtained via S^8PT_+ starting from lithographically defined seeds separated by a distance allowing the optimal arrangement of the wire arrays

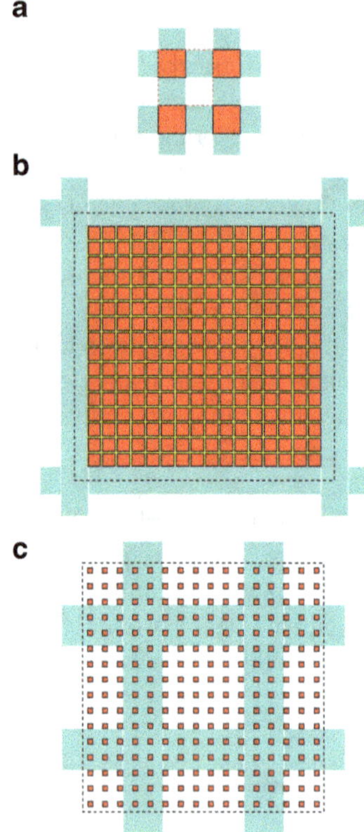

Fig. 6.4 Plan-view comparison of the crossbars obtained (**a**) crossing lithographically defined lines, (**b**) using the lithographically defined lines above as seeds for S^8PT_+, and (**c**) using the lithographically defined lines above as seeds for $S^3PT_\times$. In each structure the *square with dashed sides* denotes a unit cell suitable for the complete surface tiling

(c) A 16×16 crossbar obtained via $S^3PT_\times$ starting from lithographically defined seeds separated by a distance satisfying (5.8)

The figure has been drawn in the following hypotheses:

- The lithographic lines in (a) and (b) have width at the current limit for large-volume production, say $W = 65$ nm.
- The height loss τ is such that the maximum number of repetitions in the additive route is 8, and the sublithographic pitch is the same as shown in Fig. 5.5.
- The lithographic width of (c) is chosen to allow the minimum pitch to be consistent with the one obtained with the additive route ($W = 100$ nm), thus producing sublithographic wires with width (12.5 nm) that has been proved to be producible [124] in this way.

For $n \geq 3$, this comparison is so favorable to $S^nPT_\times$ to suggest its practical application. The following factors, however, would make S^nPT_+ preferable to $S^nPT_\times$:

1. If addressing the wires defining the crossbar are also used as addressing lines, they cannot run along the entire plane; their interruption for addressing reduces the available area.
2. In $S^nPT_\times$ all wires are produced collectively, and have the same height and material characteristics. On the contrary, in S^nPT_+ the wires are produced sequentially, each SPT_+ repetition produces wires of decreasing height, and the characteristics of the material (or even the materials themselves) may vary in a controlled way from one cycle to another.

Whereas at the first glance all the features of the second item may seem detrimental, the analysis of the Sect. 6.3 has clarified that they may be usefully exploited for crosspoint addressing.

Deciding which route, between the additive and multiplicative ones, is actually more convenient for the preparation of hybrid devices, depends on the particular application and circuit architecture. In fact, the multiplicative route (certainly less demanding for what concerns the preparation of the crossbar but more expensive for what concerns the litho-to-nano link) is presumably suitable for RAMs; on the contrary, the additive route (more demanding for the crossbar preparation, but much heavier for the elaboration of the signal) seems consistent with nonvolatile memories. Assuming that the entire crossbar is organized in modules with the same bits, different process architectures result in different numbers of contacts, as shown in Table 6.1. Additional process simplifications may be hypothesized (for instance, with the deposition of a multislab as sketched in Fig. 5.9).

6.5 Not Only Crossbars

The combination of techniques of sublithographic definition (like NIL or S^nPT) with demultiplexing techniques makes the nanoworld exploitable. Although the attention has so far been focused on the crossbar only, this combination can be used for other applications too.

Table 6.1 Number of SPT cycles, number of contacts, and size of the buffer necessary for extracting the information from crossbar module of n^2 bits

Process architecture	Cycles	Contacts	Buffer
Additive	$2 \times (1/2)n$	2×3	n^2
Multiplicative	$2 \times \log_2(n)$	$2 \times n$	1
Mixed	$(1/2)n + \log_2(n)$	$3 + n$	n

In this section, other potential applications will be considered: supercapacitors, photoluminescent nanosheets, and nanowire arrays as Seebeck generators.

6.5.1 *Supercapacitors*

In the planar technology, the capacitor is produced using two faced parallel metal plates separated by an insulator of dielectric constant K_{ins} and thickness t_{ins}. For any assigned technology (defined by K_{ins} and t_{ins}) the capacity is controlled by the faced area A. One of the major limits of the planar technology is its incapacity to produce capacitors with capacity greater than the chip capacity. A fall-out of the technological shift of the FET insulator, from SiO_2 to high-K dielectrics, will be the preparation of capacitors with correspondingly larger capacity.

The ability of producing silicon nanowires separated by insulators of thickness about t_{ins} and to connect them as required for the preparation of an interdigitated capacitor helps to overcome the maximum limit. Consider rectangle of area $L \times W$ (and thus suitable for the planar-technology preparation of a capacitor with capacity $\epsilon_0 K_{ins} LW / t_{ins}$, with ϵ_0 the vacuum permittivity. Transforming this zone in a trench of depth H and filling it with a wire array of length L with pitch p produces a total area of $2HLW/p$. If they are connected to form an interdigitated capacitor, its capacity will be given as $\epsilon_0 K_{ins}(LW/t_{ins})(H/p)$. In other words, the use of sub-lithographic wire arrays for the preparation of interdigitated capacitor results in an increase of the capacity by a factor of H/p. For $H = 3\,\mu m$ and $p = 30\,nm$, the new technology would produce a magnification of the capacity by a factor of 10^2.

6.5.2 *Photoluminescent Nanosheets*

The discovery by Canham [128] of the strong visible photoluminescence (VPL) of porous silicon (π-Si) has opened the interest toward the application of silicon in optoelectronics. Since the morphology of π-Si is characterized by columnar *nano*wires, its VPL has been attributed to a quantum size effect – the confinement of an exciton in a region with size comparable with its diameter, results in an increase of the gap and consequently in the emission of light with reduced wavelength. This interpretation has been a matter of intense discussion [142], eventually culminating in the now generally accepted smart quantum confinement (SQC) model [143]. In this model the emission energy depends not only on the quantum confinement (QC) in a structure of reduced dimensionality and nanometer size, but also on the chemical termination of the nanowire surface. Surface states have been shown to act as recombination centers for those excitons created by the exciting radiation in the wide-bandgap region.

So far, the VPL has been observed not only in silicon nanowires, but also in nanodots [144, 145] and (though to a lesser extent) in nanosieves, where the confinement is attributed to a kind of Anderson localization by voids [146].

From this situation, it seems that topology does not affect the VPL significantly: denoting with D the dimensionality of the system, one has $D = 0$ for nanodots, $D = 1$ for nanowires, and $D = 3$ for nanosieves; nanodots and nanowires are simply connected systems, whereas nanosieves are multiply connected. Rather, what seems to control the phenomenon, in addition to surface terminations, is simply the size – the VPL appears when the exciton is confined in regions of size around 3 nm.

If this conclusion is correct, two-dimensional nanosheets should also display VPL. Preliminary evidence has been put forward in recent literature, confirming such an expectation. In almost all cases (with the remarkable exception of that reported in [147]) the nanosheets were prepared by controlled oxidation of silicon-on-insulator (SOI) wafers [148–152]; in all papers but [152], the VPL was attributed to quantum confinement (but [152] attributed it to a $Si–SiO_2$ interface effect); in no case it was possible to separate the contributions of Si nanosheets and SiO_2 from VPL; in all cases but [147] the amount of VPL matter was limited to one nanosheet parallel to the surface.

Because of the considerations of Sect. 6.5.1, an array of nanosheets perpendicularly oriented with respect to the surface can be prepared by filling a trench of depth H defined over a zone of area $L \times W$ via conformal deposition of a multilayered film $Si–SiO_2|Si–SiO_2|Si-SiO_2|,\ldots$ followed by its directional etching up to the appearance of the substrate (as in Fig. 5.9).

6.5.3 Nanowire Arrays as Seebeck Generators

The Seebeck effect is the direct conversion of temperature differences to electric voltage. It can be exploited for the direct transformation of heat into electrical energy. Its potential interest is due to the fact that it does not involve any mechanical intermediate and is able to transform even low-temperature differences (although in the full respect of the Second Law of thermodynamics). The practical application of the Seebeck effect has however revealed difficulties because only expensive materials have Seebeck coefficients sufficiently large for practical exploitation [153].

However, it has recently been shown that silicon nanowires with width around 20 nm have Seebeck coefficient sufficiently high to hypothesize their use for energy conversion [154, 155].

The reason for this increase is attributed to the fact that surfaces (rugged surfaces, in particular) are able to reduce thermal conduction significantly at a distance but not electrical conduction along the wire.

Of course, the lithographic production of nanowires would make them unsuitable for practical applications, but both IL and S^nPT (and especially S^nPT$_x$) seem able to overcome the economic limits.

Chapter 7
Functional Molecules

Problem 3 (Design and synthesis of functional molecules) Designing and synthesizing bistable molecules, reconfigurable via electrical stimuli.

In most cases, the current I flowing through a single (conducting) molecule is of the order of 1 nA for an applied voltage V of the order of 1 V (thus with $V/I \approx 10^9 \, \Omega$). Clearly enough, one can speak of "resistance" only in the case of linear V–I characteristics. Although that is not the case of molecules, the conduction properties of molecules will nonetheless be referred to in terms of "effective resistance" V/I at the considered value of voltage, because it is just this quantity (rather than the derivative $\mathrm{d}V/\mathrm{d}I$ – the differential resistance) that is of major interest for practical applications.

Which is the conduction mechanism of, or inside, a molecule is likely to be an ill-posed question, because in many cases the conduction is controlled by the contacts. The role of the molecule–electrode interface has been considered in large detail for SAMs on noble or near-noble metals [156]; its role in single-molecule transistors was considered in [157]; the factors that may affect the interface between metal electrode and molecule in practical conditions has recently been discussed by Zhitenev et al. [104], while an extended review is given in [158]. The electrical properties of the silicon–organic interface are much less known.

7.1 The Molecule as a One-Dimensional Wire

The simplest case one can consider is that of a one-dimensional wire, linking two contacts at different electric potentials, of length much shorter than the mean free path for collisions against phonons populating the wire. For this situation the electron current I flowing along the wire is given by

$$I = e \int_0^{eV} g(E)v(E)\mathrm{d}E, \tag{7.1}$$

where $v(E)$ is the electron group velocity, $g(E)$ is the state density, and V is the applied voltage. For the free particle $v(E) = (2E/m)^{-1/2}$; taking into account that for a one dimensional system $g(E) = (m/\pi\hbar)(2mE)^{1/2}$, (7.1) gives

$$I = (2e^2/h)V. \tag{7.2}$$

7.1.1 The Role of Contacts: Landauer Resistance

From the formal point of view, (7.2) is nothing but the Ohm law, where the constant of proportionality between voltage and current, $h/2e^2$, is quantized, $h/2e^2 = 12.91\,\mathrm{k\Omega}$. This quantity is usually referred to as resistance quantum (or Landauer resistance).

The analogy between (7.2) and the Ohm law is however only formal.[1] This statement follows immediately not only from the quantized value of the Landauer resistance, but also from the fact that (7.2) holds true just in the absence of dissipative collisions of electrons against scattering centers, whereas the resistance of ohmic conductors is just due to such events.

For more complex situations, where the one-dimensional system can again be represented as frictionless but with a transparency $\mathcal{T}$ for electrons impinging the wire from the contact at lower potential, (7.2) becomes [167, 168]

$$I = \frac{2e^2}{h}\frac{\mathcal{T}}{1-\mathcal{T}}V. \tag{7.3}$$

Assuming the conduction along one-dimensional wires as a model for molecular conduction, the Landauer resistance (about $10^4\,\Omega$) is not the limiting factor to the current flowing through molecules.

7.1.2 Barrier Transparency

The essentially one-dimensional (1D) nature of the conduction along molecules allows the determination of $\mathcal{T}$ with Wentzel–Kramers–Brillouin (WKB)'s

---

[1] That linear I–V characteristics do not necessarily imply ohmic conduction is known also from the behavior of almost ideal silicon p–n junctions. Carefully prepared, well gettered [159–164] abrupt silicon p–n junctions, displaying an ideal behavior under forward bias [163], manifest linear I–V characteristics under reverse bias [164, 165]. This behavior may be explained assuming the presence in the depletion region of shallow donor–acceptor pairs. If they are sufficiently close to one another (as twins) they cannot act as electron–hole recombination centers [166]; being shallow centers, their generation rate is strongly sensitive to the applied voltage (Poole–Frenkel effect) and varies with $|V|^{1/2}$; if they are uniformly distributed in the depletion layer, their amount varies as $|V|^{1/2}$, thus producing a seemingly ohmic behavior.

quasiclassical approximation. Denoting with $U(x)$ denote the potential energy along x, the regions $\{x : U(x) > E\}$ are classically forbidden to any particle with energy E. From the quantum mechanical point of view, on the contrary, such regions can be explored by the particle and any potential barrier has a non-null transparency. According to WKB, the transparency of any simply connected, classically forbidden, region is given by

$$T = \exp\left(-2\int_{x_1}^{x_2} \sqrt{\frac{2m}{\hbar}(U(x) - E)}\mathrm{d}x\right) \tag{7.4}$$

where x_1 and x_2 are the boundaries of the region: $U(x_1) = U(x_2) = E$ [169, Sect. 3.3.2].

In the absence of other information, the potential is often assumed to be constant inside the classically forbidden region, $U(x) = U_0(> E)$ for $x_1 < x < x_2$. In this case (7.4) gives a transparency T_0 decreasing exponentially with the length L of the forbidden region ($L = x_2 - x_1$):

$$T_0 = \exp\left(-2\sqrt{\frac{2m}{\hbar}(U_0 - E)}L.\right) \tag{7.5}$$

For this potential, the Schrödinger equation can be solved exactly [169, Sect. 3.3.1] and the resulting transparency coincides with the WKB approximation.

Any external potential $U^{\mathrm{ext}}(x)$ superimposes additively to the internal one. T thus depends on $U^{\mathrm{ext}}(x)$, rendering the characteristic nonlinear. Seemingly ohmic characteristics may be observed only in the limit of extremely weak field, for which $U^{\mathrm{ext}}(x)$ is negligible with respect to the internal potential.

Otherwise, the characteristic depends on $U^{\mathrm{ext}}(x)$. Two limiting cases are considered: direct tunneling (for which the external field is a perturbation of the internal field of strength however insufficient to modify the extension of the forbidden region) and field emission (for which the external field is a strong perturbation of the internal field able to modify the extension of the forbidden region). In both cases the internal potential will be assumed to be constant with x.

Direct Tunneling

The situation considered here is that of an electron feeling a field obtained by superimposing a uniform electric field $\mathcal{E}$ on a forbidden region extending from 0 to L: $U^{\mathrm{ext}}(x) = -e\mathcal{E}x$. The electric field is not so strong as to modify the extension of the forbidden region, i.e., $e\mathcal{E}L < U_0 - E$. In this case one has

$$T = \exp\left(-2\int_0^L \sqrt{\frac{2m}{\hbar}(U_0 - E - e\mathcal{E}x)}\mathrm{d}x\right). \tag{7.6}$$

Considering $e\mathcal{E}x$ as a perturbation,

$$\sqrt{U_0 - E - e\mathcal{E}x} = \sqrt{(U_0 - E)}\sqrt{1 - \frac{e\mathcal{E}x}{U_0 - E}},$$

and defining

$$\epsilon(x) = e\mathcal{E}x/(U_0 - E),$$

a Maclaurin expansion of the square root gives

$$\epsilon(x) \ll 1 \Longrightarrow \sqrt{U_0 - E - e\mathcal{E}x} \simeq \sqrt{(U_0 - E)}\,(1 - \epsilon(x)),$$

so that the integration of (7.6) gives

$$T = T_0^{1-(1/4)\epsilon(L)} \tag{7.7}$$

with and V being the applied voltage, $V = \mathcal{E}L$. The following two points are noted: the theory works only for $\epsilon \ll 1$; and since $T_0 < 1$ (usually $T_0 \ll 1$) (7.7) gives T increasing with the applied voltage V ($V = \mathcal{E}L$).

Field Emission

When the applied voltage is higher than $(U_0 - E)/e$, the extension of the forbidden region decreases with L and the corresponding phenomenon, field emission, is controlled uniquely by the electric field and the energy barrier at the interface. Because of this property, the molecular properties are of less importance except for their ability to control surface potentials. This phenomenon, usually known as field emission, is of paramount importance in microelectronics because it is present in the erasure of flash memories. The theory of field emission is usually based on the assumption of a triangular (rather than trapezoidal) energy barrier and leads to a transparency given by [170, Problem 169]

$$T = \exp\left(-\frac{4}{3}\frac{\sqrt{2m}}{\hbar e\mathcal{E}}(U_0 - E)^{3/2}\right).$$

Using this transparency for the determination of the field emission from metal surfaces leads to a current density J varying with the surface field $\mathcal{E}$ as

$$J = A\frac{\mathcal{E}^2}{\Phi}\exp\left(-\frac{4}{3}\frac{\sqrt{2m}}{\hbar e}\frac{\Phi^{3/2}}{\mathcal{E}}\right), \tag{7.8}$$

with A a (universal) constant and Φ the work function. Equation (7.8) is known as Fowler–Nordheim law. Although it is widely used in the description of field emission from silicon (however considering A as an adjustable parameter), it is well ascertained that it underestimates the emitted current and that it may be accounted for considering the image force only [170, Problem 170]. The validity of (7.8) for field emission from silicon could be due to the fact that almost all increasing I–V characteristics are linearized in a Fowler–Nordheim plot ($\ln(I/V^2)$ vs. V^{-1}).

Of course, describing the conduction along molecules simply in terms of Landauer resistance and barrier transparency, thus ignoring the properties of the molecule and interface, is an oversimplification. Understanding molecular conduction requires the specification of those properties.

7.2 Conduction Along Alkanes

The simplest case one can consider is perhaps that of covalently bonded alkane at single-crystalline silicon surfaces.

The most advanced conclusions about this system are perhaps those of Cahen and coworkers for alkyl-terminated n-type (1 1 1) Si. In a series of papers that collaboration reached the conclusion that the electron transport through Si–C bound alkyl chains (SiC_nH_{2n+1}, $n = 12, 14, 16$, or 18) is characterized by two distinct types of barriers, each dominating in a different voltage range:

- At low voltage, the current depends strongly on temperature but not on molecular length, suggesting transport by thermionic emission over a barrier in the silicon.
- At higher voltage, the current decreases exponentially with molecular length, suggesting transport limited by tunneling through the molecules [171] (although the data should have been checked on (7.7)).

Whereas the tunneling characteristics appear rather insensitive to the quality of the interface, the thermionic current is strongly depressed by low-quality monolayers [172]. How it is difficult to control the superficial state of silicon, even in an industrial environment, is understood from [173].

Although the conduction along alkyl chains has been investigated (in self-assembled monolayers too [174, 175] as well as anchored to gold electrodes via amine groups [176]), their use as memory elements in crossbars seems unlikely because of the low molecular conductance and the absence of metastable states with different conductance.

Rather, the higher conductance of π-conjugated chains make them more suitable for applications in molecular electronics. Electron transport through π-conjugated chains linked to the silicon surface via Si–C bonds was investigated in [177].

7.3 Switchable π-Conjugated Molecules

The particular interest toward π-conjugated molecules is due to the fact that the separation between the highest occupied molecular orbital (HOMO) and the lowest unoccupied molecular orbital (LUMO) depends strongly on the overlapping of neighboring π orbitals, in turn related to the planarity of the molecule and to side groups. In many cases, the band gap is relatively narrow (say, around 1 eV), the molecules can be doped to assume an n- or p-type character, and electron or hole mobility is relatively high.

If the complete π conjugation of a molecule is broken at a certain site, the molecular conductance decreases and is controlled by the difficulty in transferring the electron from one π-conjugated fragment to the other. By describing a completely π-conjugated molecule as an alternate sequence of single and double bonds, the conjugation can be destroyed by any reaction able to cleave a π bond. It is however to be noted that this view is an oversimplification: Fig. 7.1 shows how an alternate sequence of single and double bonds that may be destroyed by a suitable chemical stimulus; in the considered example, however, the complete π conjugation is not destroyed because the oxidized carbon atoms maintain sp^2 hybridization. In such cases, the change of conductance is better explained in terms of molecular rotation or overlap of molecular orbitals.

The use of chemical stimuli, though of primary interest for sensors, is however useless (and potentially dangerous) for electron devices. Molecular electronics is instead interested in molecules undergoing change of conductance under electrical activation. The first pivotal result in this field was achieved by the collaboration of [12]: They reported that the molecule shown in Fig. 7.2 (a thiol-terminated nitroaniline) undergoes a dramatic increase of conductance in a narrow voltage range

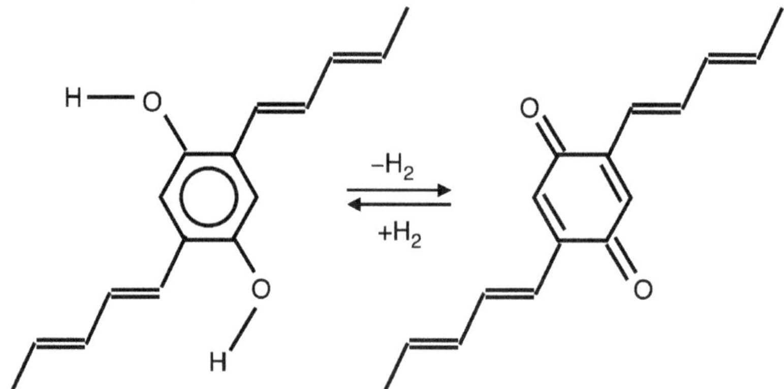

Fig. 7.1 How a sequence of alternate single and double bonds in a π-conjugated molecular wire may be broken by oxidation: (*left*), two poly-acetylene chains linked by a hydroquinone group form a semiconducting π-conjugated wire; (*right*), the quinone group, resulting from dehydrogenation of the hydroquinone group

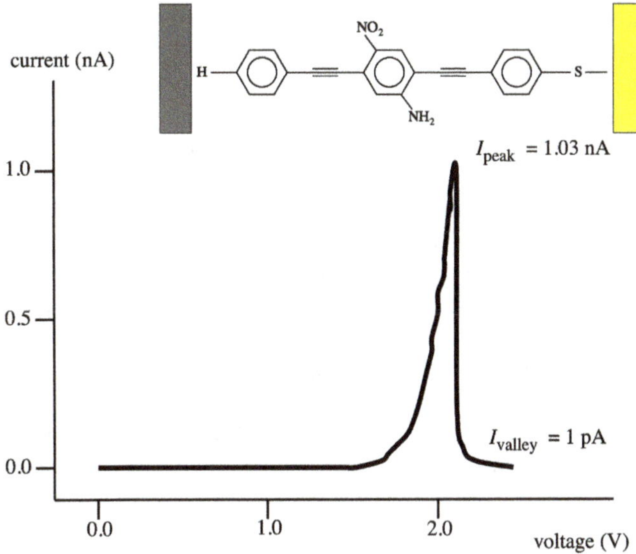

Fig. 7.2 Current–voltage characteristics of thiol-terminated nitroaniline π-conjugated molecule embedded between metal electrodes (after J. Chen et al., *Science* **286**, 1550 (1999))

at a temperature of 60 K (the current I flowing along one molecule ranging from about 1 pA at for $V < 1$ V to 1 nA at about 2V̇).[2] This situation allows the assignment of two conduction states, OFF and ON, with $I^{ON}/I^{OFF} = 10^3$. This dramatic change of conductance may be explained, advocating that the injection of an electron into the molecular orbitals of the nitroamine group leads to the generation of molecular orbitals with maximum overlap and thus maximum opportunities for conduction [178].

This explanation is not limited to the molecule of Fig. 7.2: Consider a partially π-conjugated molecule, containing a donor–acceptor pair D–A, so formed as to undergo complete π conjugation after ionization of the pair, D–A $\longrightarrow$ D$^+$–A$^-$. Molecules of this kind can be excited to form an ionic pair D$^+$–A$^-$ via the application of a strong electric field and the molecule evolves into an energy eigenstate of complete π conjugation via nuclear relaxation. Condition for this occurrence is that the D$^+$–A$^-$ distance is larger than the decoherence length. The relaxed molecule is in a metastable configuration and can be brought to the original state either via thermal excitation or by the application of a reverse electric field. Due to relaxation effects, the reverse field is in general different from the direct field thus generating the hysteresis required for the molecule to operate as a memory element.

---

[2] A hypothetical molecule that exhibited at room temperature current–voltage characteristics like that sketched in Fig. 7.2 would provide a candidate for the resonant diode considered at the end of Sect. 6.3.

The breakdown of the π conjugation may occur via the modification of a single site, rather than of a pair. Consider, for instance, a π-conjugated molecule containing a 4,4′-bipyridinium core, where two ionized nitrogen atoms, in sp^2 hybridization, form the opposite vertices of two aromatic rings and their charge is counterbalanced by a strong anion (like $[PF_6]^-$). Each nitrogen atom of the bipyridinium core can be reduced, attaining in this way an sp^3 hybridization and hence a tetrahedral conformation. This destroys the planar structure of the molecule, thus being responsible for a loss of conductance along the molecule. In so doing, however, the counterion is no longer kept in the neighborhood of the considered nitrogen atom and may therefore be lost. If this is the case, cycling the nitrogen between oxidized and neutral state becomes impossible and so also its use in memories. To avoid this loss, the counterion should be covalently linked to the chain by a flexible (e.g., alkyl) chain (an early exploration of this possibility goes as far back as to late 1980s [179]); the existence of a neutral Lewis acid able to dock the fluctuating anion on another side chain would result in a new equilibrium configuration eventually responsible for bistability.

7.4 Molecules Exhibiting Superexchange Conduction

The strong interest for π-conjugated molecules is due to the fact that many of them have relatively large conductance. Most of the other organic molecules are instead characterized by the absence of any, even approximate, translational invariance and are therefore very poor conductors. This situation is characterized by largely separated HOMO and LUMO levels, with deep HOMO and shallow LUMO. In these conditions, the Fermi level of common metals and semiconductors (irrespective of dopant concentration) lies in the gap between molecular LUMO and HOMO and is separated from them by an energy appreciably larger than $k_B T$. In these conditions, the conduction along the molecule occurs via tunneling and its small conductance is controlled by the transparency of the energy barrier separating the contacts. In these conditions, the molecule can be considered to be in an OFF state.

The conductance can significantly be increased (even by several order of magnitude) by simply introducing in the HOMO–LUMO gap, allowed energy levels aligned (to within a few $k_B T$) with the Fermi level of the emitting contact. If the newly generated levels are few (possibly 1) and spread in energy over an interval of the order of $k_B T$, the molecule will be brought into an ON conduction state only for applied voltages in a restricted interval (thus displaying a behavior similar to that reported in [12]); on the contrary, the generation of many levels spread over an energy interval much larger than $k_B T$ will produce a switch in an ON state for all voltages larger than a critical one.

For applications in molecular electronics it is required that the said levels are generated under electrical activation; that immediately leads to search for levels generated via a redox process involving electron transfer from the molecule to one electrode or vice versa.

Let the functional molecule be characterized by a closed-shell configuration and let it be assumed that the redox process involves one electron only with the formation of an unpaired state $E^{\bullet+}$ with an allowed energy level roughly aligned with the Fermi level of the emitting contact (although the unpaired center is assumed to be positively charged, mutatis mutandis the forthcoming considerations can be extended to a negatively charged center). This situation allows that the wave function of the tunneling electron (decreasing exponentially in the forbidden region) reacquires an oscillatory behavior ("refreshes") in $E^{\bullet+}$ before decreasing again exponentially on tunneling from the $E^{\bullet+}$ region to the other contact. It is just the refresh of the wave function in the region of $E^{\bullet+}$ that allows an (often dramatic) increase of the tunneling rate.

Evidence for this kind of mechanism was given by Gittins et al. [13]. Rather than the π-conjugated chains considered in the previous part, long alkyl chains were attached to the bipyridium core. In these conditions, the molecule behaves as a nearly perfect insulator with negligible current. The reduction of one nitrogen atom in the core, however, produced the raise of an allowed level in the gap thus being responsible for a dramatic increase of current; the reduction of the second nitrogen atom (restoring a closed-shell configuration via filling of π^* orbitals) restored the original OFF conduction state. The above experiment was carried out under electrochemical conditions, with the molecular configuration stabilized by an external potential. Whereas under these conditions one can ignore the fate of the counterion (that goes into the solution), the use of such a molecular configuration in solid state would require its covalent insertion in the molecule as discussed in the previous part.

One major difficulty with this mechanism comes from the fact that a tunneling electron in a nearly stationary electronic state of the region of $E_*^{\bullet+}$ is just an electronically excited state E_*^0 of neutral ground state E^0:

$$E_*^{\bullet+} + e^- \longrightarrow E_*^0. \tag{7.9}$$

The excited state can decay not only along the desired channel, i.e., releasing the electron to the other contact via tunnel effect,

$$E_*^0 \xrightarrow{\tau_{\text{tunnel}}} E_*^{\bullet+} + e^-, \tag{7.10}$$

but also changing its conformation (nuclear relaxation) to attain the ground state E^0 of the nonconducting neutral molecule:

$$E_*^0 \xrightarrow{\tau_{\text{rel}}} E^0, \tag{7.11}$$

where τ_{tunnel} and τ_{rel} are the corresponding lifetimes. This second channel destroys the information.

The prevalence of one phenomenon on the other is essentially controlled by the relative lifetimes τ_{rel} and τ_{tunnel}; the former can be evaluated via ab initio molecular dynamics, while the latter (of difficult evaluation [167, 168, 180, 181]) is expected to be very high for low applied fields (thus rendering the tunneling rate negligible). This situation would tend to destroy such metastable states under READ condition.

The possibility of using metastable states for nonvolatile memories is thus facilitated by either increasing the lifetime τ_{rel}, or avoiding the passage of electrons in the region $E_*^{\bullet+}$: The first approach was suggested by the group of Heath and Stoddart with the use of rotaxanes [96, 100, 102, 107, 157], where the lifetime is increased because reaction (7.11) is made difficult by the anchorage of macrocyclic rings on a second site of the conducting chain. The second approach was suggested by Cerofolini and Mascolo [126] with the use of molecules where the redox center is not along the conductive path, but rather on a side, and able to vary the transparency of the barrier on pure electrostatic bases. It is however noted that whereas there are a lot of examples of the first route, no practical demonstration has been given yet for the second route.

7.5 A Comparison of the Switching Mechanisms

The complexity of the phenomena involved in the switching of reconfigurable molecules is such that a unified understanding is still missing. In spite of that, in the authors' belief, the following rationale captures the essential phenomena occurring in π-conjugated molecules and in molecules displaying superexchange conduction. A pictorial view of the reasons why these molecules may have very different conductances is shown in Fig. 7.3.

The upper part of this figure shows the energy configurations of a π-conjugated molecule in OFF and ON states. Denote with $m + n + 2$ the number of π–σ units forming the molecule and assume that it contains a strong Brønsted acid as a side

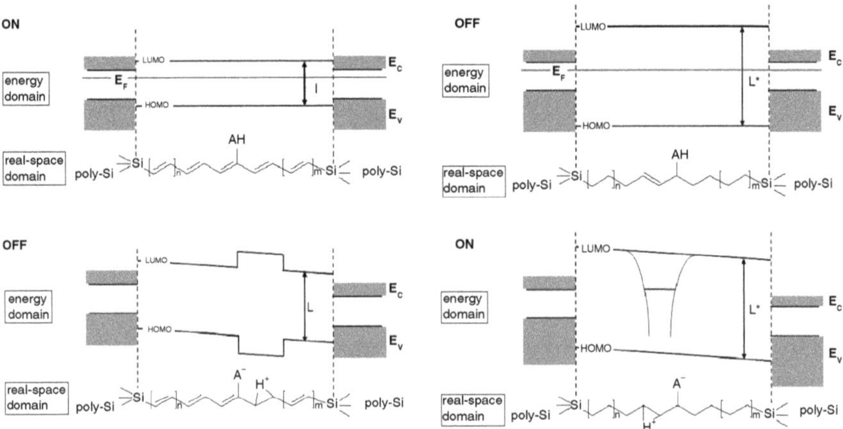

Fig. 7.3 Rationale showing the different conduction states in π-conjugated molecules (*up*) and in molecules displaying superexchange conduction (*down*)

group at unit $n + 1$ (a similar argument, however, can be repeated for Lewis acids too). The HOMO–LUMO difference is controlled by the nature of the π–σ unit and by the length $m + n + 2$ of the chain (the longer the chain, the lower the difference). If the LUMO level is close to the Fermi level of contact, electronic conduction is easy and an ON state can be assigned to this molecular configuration (right). Assume now that under the application of a strong electric field the Brønsted acid undergoes dissociation donating the proton to a neighboring π bond. After nuclear relaxation, the electric field is switched off thus leaving a molecule where the π conjugation is broken at the original $n + 2$ unit (left). In this state, the HOMO–LUMO difference increases because of the lower chain lengths of π-conjugated fragments and of the loss of π conjugation at unit $n + 2$. Because of this situation the molecule can be used to code an OFF state. The hysteresis required for memory application is essentially determined by relaxation effects and can be controlled by imparting a judicious stress.

The lower part of Fig. 7.3 shows instead a molecule displaying conduction via superexchange mechanism. The molecule is linear and formed by two chains of m and n CH_2 units, separated by a pair of double-bonded carbons and by a carbon having a strong Brønsted acid as a side group. If the conduction occurs via direct tunneling along the chain, the conductance is manifestly low and the molecular configuration can be used to code an OFF state (left). As above, assume that under the application of a strong electric field the Brønsted acid undergoes dissociation donating the proton to a neighboring π bond. After nuclear relaxation, the electric field is switched off, leaving a positively charged site around positions $n + 1$ and $n + 2$ along the chain. If this potential admits an allowed level at energy close to the fermi level of the emitting contact, the transparency of the barrier increases significantly, thus allowing the molecular configuration to code an ON conduction state (right).

Chapter 8
Grafting Functional Molecules

Problem 4 (Functionalization of sublithographic features) Adding via batch processing assigned molecules to sublithographic features.

Needless to say, strategy XB_* requires the ability to attach *organic* moieties to the silicon surface, thus forming a hybrid interface with assigned electrical properties. It is also noted that since the crosspoint regions are recessed, not only is the use of scanning probe techniques disallowed by hypothesis, but also impossible.

The needs of

- Increasing the reactivity of the silicon surface to the desired organic derivatization
- Attaining the maximum coverage degree
- Allowing the molecules to self-assemble in an ordered structure, determined uniquely by intermolecular lateral forces and steric hindrance

are generally assumed to require a prior homogeneous termination of the surface. The nature of terminal species (halogen, hydrogen, hydroxyl, or others) is fixed by the grafting route.

8.1 Silicon and Its Surfaces

The first organosilicon compound, Et_4Si, was prepared over 100 years ago by Friedel and Crafts. Silicones are organosilicon compounds with $+Si\!-\!O+_n$ backbone and organic (usually alkyl) side chains. Although silicones were recognized to be extremely useful polymers already in the 1930s, a commercially viable route to these materials was independently discovered by Eugene Rochow (at General Electrics) and Richard Müller (at VEB Silikon-Chemie) in the 1940s – the "direct process" for synthesis of organochlorosilanes.[1] The organosilicon field has since then grown to a branch of chemical science currently in full flower with annual worldwide commercial sales in the range of several billion dollars.

---

[1] The short historical note given above is taken from [182, pp. 382–383].

All silicon chemistry starts from the production of elemental silicon, which is produced by reduction with coal of silica

$$SiO_2 + C \xrightarrow{\Delta} Si + CO_2.$$

This process is highly endothermic, and practical only in an electric furnace; the unrecoverable electric energy spent in this step is a major contributor to the cost of organosilicon compounds.[2] From elemental silicon, there are three major routes to the C–Si bond, and all of them involve energy expenditure:

Direct process involves the oxidative addition of chloroalkane RCl to elemental silicon:

$$\overset{\displaystyle \diagdown \quad \diagup}{\underset{\diagup \quad \diagdown}{-Si-Si-}} + RCl \xrightarrow{\Delta,M} \overset{\displaystyle \diagdown}{\underset{\diagup}{-Si-Cl}} + R\overset{\diagup}{\underset{\diagdown}{-Si-}} \tag{8.1}$$

with M being a transition metal catalyst (like copper, nickel, or tin). The reaction proceeds with the formation of all possible organochlorosilanes. Among the several reactions, the one of largest industrial interest is the following:

$$Si + 2CH_3Cl \longrightarrow (CH_3)_2SiCl_2 \tag{8.2}$$

because $(CH_3)_2SiCl_2$ is the monomer involved in the silicone synthesis. Of all possible catalysts, a high selectivity toward (8.2) is achieved with copper catalysts only, whereas the distribution of monomers obtained with other metals disfavors $(CH_3)_2SiCl_2$; the reasons supporting copper as catalyst are discussed in [182, p. 383].

Hydrosilylation involves first the oxidative addition of HCl to elemental silicon to form Cl_3SiH (another direct process), and then the reaction trichlorosilane with an alkene:

$$Si + 3HCl \xrightarrow{\Delta} Cl_3SiH + H_2 \tag{8.3}$$

$$Cl_3SiH + CH_2{=}CHR \longrightarrow Cl_3Si{-}CH_2CH_2R \tag{8.4}$$

Reaction (8.3) is used for the preparation of the precursor of single crystalline silicon; at room temperature (8.4) requires catalysts (Lewis acids or platinum catalyst), but occurs also under thermal activation only.

Grignard process involves first the reaction of elemental silicon with Cl_2 to form $SiCl_4$, and then the reaction tetrachlorosilane with a Grignard reagent RMgX (with X denoting a halogen):

$$Si + 2Cl_2 \xrightarrow{\Delta} SiCl_4 \tag{8.5}$$

$$SiCl_4 + 4RMgX \longrightarrow SiR_4 + 4ClX \tag{8.6}$$

---

[2] A direct path from sand to silicones, with the attractive stoichiometry $2R{-}H + SiO_2 \longrightarrow R_2Si(OH)_2$, is still a dream.

8.1.1 Silicon Chemistry

Defining strategies for silicon functionalization requires a prior knowledge of its chemistry.

All the examples considered so far require the ability to bind organic or organometallic molecules to silicon surfaces forming an organic layer with assigned electrical properties. Silicon derivatization usually requires a preliminary step leading to an as far as possible homogeneous termination of the surface with species such as halogens, hydrogen, and hydroxyl. Moreover, the steric constraints imposed by the surface exposed to the derivatizing agent imply size and shape selectivity. This item is addressed to review the state of the art about the derivatization with organic molecules of both single-crystalline silicon [and especially of its (1 1 1) and (1 0 0) surfaces, of interest for IC processing or molecular electronics] and of porous silicon (of interest not only for sensors, but also for optical devices).

Even though there are some analogies between the chemistries of carbon and silicon, the differences are prevailing. Similar to carbon, most of silicon compounds are consistent with sp^3 hybridization of outer atomic orbitals of silicon, whose atom in ground state has electronic configuration is $3s^2 3p^2 3d^0$. Instead, dissimilar from carbon:

- Numerous penta- or hexacoordinated compounds of silicon are known
- sp^2 and sp^1 hybridizations are observed in very few compounds
- Silicon has poor tendency toward catenation

Penta- and hexacoordinated compounds are usually, but not uniquely, observed when silicon is bonded to very electronegative elements, as in $[SiF_6]^{2-}$. These compounds might easily be explained in terms of lone-electron-pair donation to the empty d orbital of silicon, but the participation of this orbital in the formation of the bonds is still matter of scholarly disagreement [183].

The negligible tendency of silicon toward the formation of π bonds is determined by the fact that the stability of the Si=Si double bond is lower than the stability of the Si–Si single bond (see Table 8.1 [184]); silenes are thus unstable toward spontaneous decomposition unless the π bonds are protected against internal rearrangements by suitably bulky side groups [185].

The Si–Si σ bond behaves similarly to the π bond in olefins [183]. This analogy is sustained not only by the corresponding bond energies, but also from the following observations:

- Disilanes undergo electrophilic cleavage by the same reagents which add to olefins by cleavage of the π bond:

$$R_3Si\text{–}SiR_3 + Br_2 \longrightarrow 2R_3SiBr,$$
$$R_2C\text{=}CR_2 + Br_2 \longrightarrow BrR_2C\text{–}CR_2Br.$$

- The Si–Si bond forms dative complexes with Lewis acids.

Table 8.1 Selected bond dissociation energies E_{diss} and bond energies E_b involving silicon

Bond	E_{diss} (eV)	E_b (eV)
Si–H	3.6–4.5	
Si–C	3.9–4.1	3.3
Si–Si	3.2–3.4	2.3
Si–F	6.6–7.2	
Si–O	5.3–5.9	4.8
Si–N	4.2–4.7	
$H_2Si{=}CH_2$		1.6[a]
$H_2Si{=}SiH_2$		1.2[a]
$H_2Si{=}O$		2.6[a]
$H_2C{=}CH_2$		2.8[a]
$H_2C{=}O$		3.25[a]

[a] Of π bond only

- The Si–Si bond undergoes ionization with formation of a radical ion $R_3Si^{\bullet}\cdots$ $^{+}SiR_3$, where the complex is not destroyed, as it happens for olefins (e.g., $H_2C^{\bullet}{-}^{+}CH_2$).

The obvious difference is that while the homolytic cleavage of the π bond in C=C does not demolish the molecule (unless the reagent does not add additively to the σ bond too), the homolytic cleavage of the σ bond in Si–Si eventually results in the formation of monomers. This process is at the basis of the most important silicon reaction, the direct reaction:

$$\overset{\diagdown}{\underset{\diagup}{-Si}}{-}\overset{\diagup}{\underset{\diagdown}{Si}}{-} + RX \longrightarrow \overset{\diagdown}{\underset{\diagup}{-Si}}{-}R + X{-}\overset{\diagup}{\underset{\diagdown}{Si}}{-}$$

with X being a halogen (typically $X = Cl$) and $R = H$, CH_3, C_2H_5, C_6H_5, etc. (of which (8.1) is a special case).

This analogy clarifies the reason why, contrarily to carbon, silicon has poor tendency toward catenation. Although silanes Si_nH_{2n+2} are formally the silicon analogues of alkanes, taking into consideration their chemical properties they are the analogues of cumulenes, chemical compounds with two or more consecutive double bonds. Long cumulenes are known to be unstable; correspondingly Si_nH_{2n+2} decompose spontaneously to silicon and H_2 for $n > 3$.

8.1.2 The Role of Surfaces

Before considering in detail the various processes for silicon derivatization, the role of the underlying solid on surface reactivity will be addressed. Compared to gas- or liquid-phase reactions, the surface introduces several modifications. In fact, imagining of insulating a small portion of the surface from the remainder:

(A) The reactivity of any group therein toward a reactant in gas or liquid phase is
     expected to be the same as that of the considered group in the considered phase,

provided that

(B) The frontier molecular orbitals are replaced by valence and conduction bands
     (possibly with population controlled by doping of the solid).
(C) The steric constraint of the surface (which can in turn be considered as an
     extremely bulky group holding approximately half a space) is considered.

Feature (A) explains how gas-phase chemistry can be imported into surface chemistry to predict on analogical bases surface reactivity; features (B) and (C) define the
limits of the analogy. Feature (B) is extremely important in redox reactions, where
an electron transfer from/to surface occurs and the involved energy levels play a fundamental role. On the contrary, steric constraint plays certainly an important role in
any type of on-surface reaction; it forbids free geometrical arrangement of intermediate species and final adduct, and reduces the solvent effect occupying a part of the
solvation sphere. Of course, steric constraint becomes critical when the reactant has
a size comparable with the short-range peak-to-valley roughness of the surface; it
is intuitively obvious that an atomically flat surface may be derivatized with a more
ordered and dense film than a rough surface.

To understand how strange the chemistry at a surface may be and to rationalize
it, consider on one side that silenes in gas phase can only be prepared by protecting the Si=Si double bond with sufficiently bulky unreactive groups, which
protect the π bond not only against other molecules but also against internal rearrangements [185, 186]; their effect is thus progressively destroyed as temperature
increases. On another side, observe that the clean, 2×1 reconstructed, (1 0 0) surface
of single crystalline silicon may be seen as a two-dimensional lattice of silene units
(for more information on silicon dimers at the 2×1 (1 0 0) Si_2, see [187]) stabilized
by the underlying entire crystal – an extremely bulky group. Since the silene bonds
are accessible to external molecules, they are extremely reactive (and indeed they
can be preserved in ultra-high vacua only), but they are nonetheless stable at very
high temperature too (indeed, they are formed, for instance, by thermal decomposition of silanic bonds at hydrogen-terminated surfaces at temperature higher than
approximately 750°C).

8.1.3 The Surface of Single-Crystalline Silicon

Many crystals can be prepared in almost ideal conditions with negligible content of
impurities and defects. Whichever of such single crystals is considered, its surfaces
can never be pictured as simple truncations of the crystal; rather, they are flawed by
several deviations from that ideal situation. The unsaturated forces originating from
the cutting eventually result in the relaxation of surface planes and in the reconstruction of the superficial lattice; moreover, the highly stressed configurations resulting

from relaxation and reconstruction are easily destroyed via reaction of the surface atoms with the outer atmosphere.

The situation of silicon is paradigmatic: Single crystal silicon with $\langle 1\,0\,0 \rangle$ orientation can be prepared with diameter as large as $30\,cm$, weight of the order of 10^5 g, maximum impurity (oxygen) content of the order of 1 ppm (controlled, and beneficial on mechanical properties), and point-like (vacancy and self-interstitials) and extended (dislocations and stacking faults) defects below detectability. Due to unavoidable misalignments during cutting, only vicinal $(1\,0\,0)$ surfaces with high Miller indices of such crystals can instead be prepared. These surfaces relax to more stable rough configurations via $(1\,0\,0)$ and $(1\,1\,1)$ faceting [alternating $(1\,0\,0)$ and $(1\,1\,1)$ facets produce $(1\,1\,3)$ orientation]; roughening can be removed via high-temperature annealing in ultra-high vacuum eventually resulting in stepped surfaces with steps with 2×1 reconstruction and length (typically of 10^{-5} cm) controlled by the cutting misalignment [188]. Such surfaces are, however, extremely reactive and react almost instantaneously with ambient air to form a heterogeneous distribution of Si–H and Si–O– centers [189–191].

Silicon wafers are commercially available with various orientations: $\langle 1\,1\,1 \rangle$, $\langle 1\,0\,0 \rangle$, $\langle 1\,1\,0 \rangle$, $\langle 1\,1\,3 \rangle$, etc. From them, only the $(1\,0\,0)$ and $(1\,1\,1)$ surfaces are currently produced and used in semiconductor silicon processing – and the $(1\,0\,0)$ surface alone covers almost all applications. For many years the quality of such a surface has been related to the following processes:

- The polishing procedure
- The cleaning process after polishing

The polishing procedure is essentially a chemomechanical process carried out with a slurry of SiO_2 nanoparticles dispersed in a basic KOH solution. The polishing not only produces surfaces with prevailing hydrogen terminations, but also containing residual oxo centers (Si–O–Si and Si–O–H) [192, 193]).

The most widely used cleaning processes were developed by RCA[3] in the mid-1960s. The RCA cleaning employs two steps:

(SC1) Standard cleaning 1, where the wafer is exposed to a hot mixture of $H_2O_{2,aq} + NH_{3,aq}$, aimed at removing organics and particles
(SC2) Standard cleaning 2, where the wafer is exposed to a hot mixture of $H_2O_{2,aq} + HCl_{aq}$, designed to remove metals

The combination of this procedure with "piranha" etching (a concentrated aqueous solution of H_2SO_4 and H_2O_2) before SC1, between SC1 and SC2 or after SC2, leaves the surface oxidized and hydroxilated. The oxo-terminated surfaces exposed to humidity at room temperature undergo further oxidation [194, 195] and the early stages of oxidation produce a chemically heterogeneous surface region where the silicon oxidation state is distributed between 0 and 4 [196]. Finally, an etching in HF_{aq} followed by a rinsing in deionized H_2O is usually considered

---

[3] RCA Corporation, founded as Radio Corporation of America, was an electronics company in existence from 1919 to 1986.

an adequate treatment achieving the surface hydrogenation [197]. The extent of residual oxidation and hydroxylation depends on HF_{aq} concentration, solution pH, concentration of O_2 dissolved in water, air humidity, and duration of exposure. Since these quantities are often known only partially, the state of the surface (the native surface) is generally unknown in a broad distribution of allowed states.

Over the last years, the attention has gradually shifted from the removal of contaminants (organics, metals, particulates) to the chemical termination of the surface after the cleaning process because it affects appreciably the behavior of the Si–SiO_2 interface upon gate oxidation in integrated-circuit processing. In the following the attention will be focused on the methods developed for the preparation of surfaces minimizing the oxo terminations.

Clean Surfaces: Reconstructed and Highly Reactive

The (1 1 1) Surface There are in principle two (1 1 1) surfaces: one is characterized by three unsaturated dangling bonds per surface atom, and the other is characterized by one dangling bond per atom perpendicularly oriented to the surface. The attention is usually limited to the second case because it occurs at the cleavage plane of silicon.

The clean unreconstructed 1×1 (1 1 1) surface is unstable and reconstructs first with the formation of 2×1 (1 1 1) dimers and eventually forming a complex 7×7 (1 1 1) structure. The passivation of the dangling bonds at the unreconstructed surface produces an almost ideal 1×1 (1 1 1) SiH surface [198].

The (1 0 0) Surface The ideal (1 0 0) surface is characterized by two dangling bonds per surface atom. This surface is unstable and admits several reconstructions. The most stable one is the 2×1 (1 0 0) reconstruction, characterized by double-bonded silicon dimers in silene configuration (the bonds being a strained σ bond and a very weak π bond). This surface is formed upon heating in an ultra-high vacuum (UHV); because of the weakness of the π bond, the 2×1 (1 0 0) Si_2 dimer reacts almost immediately with air [199]. The exposure to O_2 at room temperature results in a complex transient pattern of paramagnetically active defects lasting more than 90 min at 10^{-5} Pa. This transient behavior disappears after exposure to atmospheric pressure [200]. A similar behavior is observed also for the 7×7 (1 1 1) silicon surface [200]. The oxidation proceeds even at lower temperature. At 173 K in addition to the formation of Si–O–Si centers, O_2 is reported to form also the silanone center, $\diagdown Si{=}O$, that disappears only after annealing at about 500 K [201].

Hydrogen-Terminated Surfaces: The Search of Environmentally Stable Surfaces

The saturation of unsaturated bonds with hydrogen seems the way of minimum perturbation to impart stability to otherwise clean surfaces. The relatively high strength of the Si–H bond and its modest polarization suggested its indefinite stability (on the laboratory time scale) in air at room temperature. Actually, this common belief has recently been invalidated because the Si–H bond reacts with negligible activation energy with O_2 via the following reaction [202]:

$$Si\text{–}H + O_2 \longrightarrow Si^\bullet + HO_2^\bullet.$$

The attachment of O_2 to silanic hydrogen has, however, a sticking coefficient that a hydrogen-terminated surface appears as unchanged after an exposure to air for duration of the order of 1 day [203]. This explains why several methods for the hydrogen termination of (1 0 0) silicon have been developed. They include both the wet and dry processes:

- The wet etch with HF_{aq} under special pH conditions of a sacrificial, thermally grown oxide
- A two-step process, involving

  (a) The demolition of the native oxide by heating (at $T > 850°C$) under UHV with the formation of a clean 2×1-reconstructed surface
  (b) The subsequent exposure to controlled amounts of atomic hydrogen H

- The exposure of the native surface to H_2 at subatmospheric pressure (typically in the range 10^2–10^3 Pa) and high temperature (typically 800–1,100°C).

Hydrogen-terminated surfaces can greatly differ according to the preparation procedure, as discussed in the following paragraphs.

The Surface Resulting After Aqueous HF Etching of Sacrificially Grown SiO$_2$
The dipping of the native or thermally grown oxide in HFaq results typically in a rough and heterogeneous surface. The quality of the surface is controlled by the pH of the etching solution (in turn controlled by HF_{aq} dilution and the presence of buffering species such as NH_4F or of acids such as HCl), the amount of dissolved O_2 in the water, and the duration of rinsing. Nonetheless, the following features are usually observed quite irrespective of these details. The (1 0 0) surface resulting after HF_{aq} etch displays 1×1 reconstruction, and is characterized by a heterogeneous distribution of SiH_3, SiH_2 (prevalent) and SiH terminations [204, 205]. Inspected by scanning tunneling microscopy, HF_{aq}-etched surfaces usually display no evidence for regularity; images with atomic resolution, giving evidence for 1×1 symmetry, have been obtained only using as etchant a very acidic solution (HF:HCl = 1:19, pH < 1) [206]. Almost ideal hydrogen termination of the (1 1 1) surface is instead obtained after etching in an HF aqueous solution buffered with NH_4F or NH_3 at pH > 5 [207].

    Understanding why the silicon surface after etching in HF_{aq} has prevailing hydrogen, rather than fluorine, terminations is not trivial, especially in view of the fact that silicon and fluorine combine to form one of the most stable bonds. The first explanation of this fact was given by Ubara et al. in terms of polarization of Si–Si backbonds to Si–F bonds. Assuming that the strength of the Si–F bond derives from an electrostatic reinforcement because of electron transfer from silicon to fluorine, the ionicity of the Si–F bond polarizes the Si–Si backbonds, which allows an easy insertion of HF into the weakened Si–SiF bond. If this bond is cleaved with the addition of fluorine to the SiF site and of hydrogen to the other silicon atom, the ionicity is increased and so is the polarization of the residual backbonds. Reiterating this argument, one obtains that the formation of an SiF_n moiety becomes easier the higher is n, and that the process is concluded with the formation of a volatile molecule, SiF_4, and of hydrogen terminations at the silicon surface [208]. This scheme was investigated quantum mechanically on model molecules (such as $H_3Si–SiH_2F$ or H_3SiF) by Trucks et al. They showed that the activation energy of the process leading to two fluorine atoms attached to the same silicon atom is 1.0 eV, while the transferring of fluorine to the adjacent silicon atom needs an activation energy of 1.4 eV. Then, the latter process is not favored and the reiterated addition of fluorine on the same SiF site is confirmed. Moreover, they showed that the Si–H bond is spontaneously destroyed by reaction with HF (the reaction being exothermic by 1.3 eV), but an activation energy of 1.2 eV was calculated for this process. The difference between the activation energy for the destruction of the Si–H bond (1.2 eV) and that for the attachment of another fluorine atom to fluorinated silicon (1.0 eV) is of 0.2 eV. Although small, it is sufficient to make the time constant of the former reaction four orders of magnitude higher than that of latter reaction [209]. These calculations support therefore the opinion, first formulated by Ubara et al. that the formation of hydrogen-terminated surfaces is kinetically, rather than thermodynamically, controlled. The assumption that undissociated HF is the etching agent is, however, contradicted by the following facts:

1. In diluted aqueous solution, HF is completely dissociated:

$$HF + H_2O \longrightarrow F^- + H_3O^+.$$

   Its behavior as a weak acid is a result of the reduction of the H_3O^+ activity resulting from tight ion pairing $H_3O^+ \cdots F^-$ between the oxonium and fluoride ions (see [210, Chap. 3]).
2. In concentrated aqueous solution, the H_3O^+ activity is increased because of the formation of the hydrogen-bonded adduct $F^- \cdots HF$ via the equilibrium

$$H_3O^+ \cdots F^- + HF \longrightarrow F^- \cdots HF + H_3O^+.$$

   The concentrated aqueous solution does therefore contains undissociated HF, but its binding energy to F^- is so high ($E_{diss}[F^- \cdots HF] = 2.2$ eV; [211, p. 76]) that the adduct $F^- \cdots HF$ is considered a well-defined ion HF_2^- (see [210, Chap. 3]).
3. In gas phase (where the etching species is undissociated HF), the attack results mainly in fluorine-terminated silicon surfaces [212].

Following the criticism of [213] and using the results of high-level quantum mechanical calculations of the stability of the Si Si bond in relation to its terminations, Cerofolini thus proposed an ionic route able to account for the observed hydrogen termination resulting after attack by HF-concentrated aqueous solution to thermally oxidized silicon [214]. The route is constituted by three cycles each composed by four consecutive steps, the rate-determining one being F^- transfer from the etching solution to the coma of SiF_n termination [215].

Hydrogen-Terminated Silicon via Exposure to Atomic Hydrogen The exposure of $2 \times 1\,(1\,0\,0)\,Si_2$ to a few langmuirs of atomic hydrogen results in nearly homogeneous surfaces. The obtained surface phase depends on the silicon temperature [199]:

- The 1×1 dihydride phase, $1 \times 1\,(1\,0\,0)\,SiH_2$, is produced at 300 K
- The 3×1 mixed monohydride–dihydride phase, $3 \times 1\,(1\,0\,0)\,SiH_2(SiH)_2$, is produced at 400 K
- The 2×1 monohydride phase, $2 \times 1\,(1\,0\,0)(SiH)_2$, is produced at 600 K

The thermodynamics of the hydrogen-terminated $(1\,0\,0)$ surfaces of silicon is sketched in Fig. 8.1.

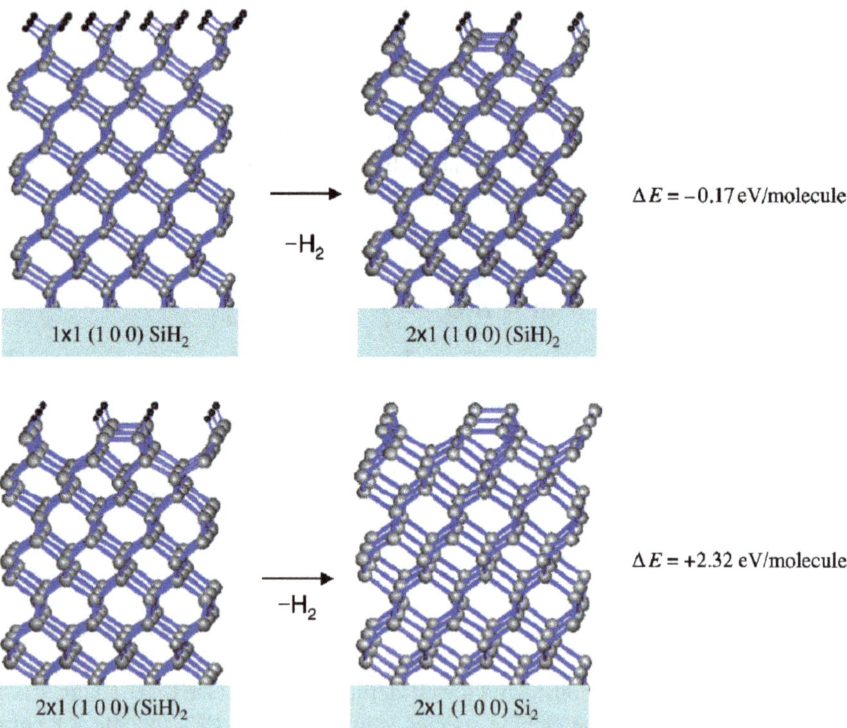

Fig. 8.1 Thermodynamics of the $(1\,0\,0)$ surface of silicon in H_2 environment (calculations carried using density functional theory within the generalized gradient approximation, after G.F. Cerofolini et al., *J. Chem. Phys.* **130**, 184702 (2009))

Hydrogen-Terminated Silicon via Exposure to Molecular Hydrogen Homogeneously hydrogen-terminated silicon is also obtained after exposing the HF_{aq} etched surface to H_2 at subatmospheric pressure in the temperature interval $800–1,100°C$. According to the operative conditions, this process results in 2×1 $(1\,0\,0)\,(SiH)_2$ [204, 216] or 1×1 $(1\,0\,0)$ SiH_2 [217]. The conditions that allow one phase to prevail on the other are not clear yet.

The Evolution in Air of Hydrogen-Terminated Silicon As already mentioned, an ideally hydrogen-terminated surface is known to be remarkably stable (on the laboratory time scale) in air. The $(1\,0\,0)$ surface resulting after HF_{aq} etching of the overlying oxide, however, not only is formed by a heterogeneous family of mono-, di-, and trihydride species, but also contains silicon–oxygen centers, their amount depending on the preparation [195]. Such oxygen-containing centers are responsible for the environmental instability of these surfaces via the oxidation of the backbonds to Si–O bonds; this process, initiated at oxygen defects and mediated by adsorbed water, proceeds with the growth of an oxide cluster, and is eventually responsible for surface roughening [195]. A microscopic model was proposed by Cerofolini and coworkers in a series of papers in terms of tunneling either of a proton from adsorbed water [218, 219] or of an electron to adsorbed oxygen [220].

Halogen-Terminated Surfaces

The great environmental stability of the Si–H bond is guaranteed more by its modest polarization than by its strength. For instance, the Si–H bond dissociation energy in $H_3Si–H$ is $92\,kcal\,mol^{-1}$ vs. an Si–F bond dissociation energy of $157\,kcal\,mol^{-1}$ in H_3SiF and an Si–Cl bond dissociation energy of $109\,kcal\,mol^{-1}$ in H_3SiCl. The Si–H bond dissociation energy is, however, nearly the same as the Si–Br one ($90\,kcal\,mol^{-1}$ in H_3SiBr) and appreciably larger than Si–I one ($71\,kcal\,mol^{-1}$ in H_3SiI) [184]. The relatively large polarization and length and the weak dissociation energy of the Si–I bond make this an interesting site for the functionalization of the silicon surface.

Halogen-terminated surfaces are prepared by high-temperature treatment of hydrogen-terminated surfaces with the corresponding halogen:

$$Si–H + X_2 \longrightarrow Si–X + HX.$$

8.1.4 The Surface of Polycrystalline Silicon

With "poly-Si" one intends a variety of materials, typically in the form of ingots, tiles or films.

Poly-Si ingots are produced via pyrolytic decomposition of $SiHCl_3$; this material is used (after crushing, melting, and crystallization) as intermediate for the growth in Czochralski pullers of single-crystalline ingots.

Poly-Si tiles are produced by melting in Bridgman furnaces the scraps remaining when the single crystalline ingots are sliced for wafer production. The polycrystalline silicon grown by the Bridgman technique is formed by large grains (with diameter in the 1–10 cm length scale) and is one of the main materials for photovoltaic applications.

Poly-Si films for microelectronics are generally produced via chemical vapor deposition from SiH_4. The typical thickness of these films is in interval 0.1–1 μm. The large interest of poly-Si films in nanoelectronics is determined by the fact that in crossbar structure at least one (but most likely two, see Chap. 4) wire array is formed by poly-Si. In poly-Si films, the average grain diameter depends on the growth temperature and ranges from the nanometer length scale (for low-temperature deposition, say $<600°C$) to the whole film thickness (for high-temperature deposition, say $>800°C$); the grain diameter increases after heat treatments. After sufficiently strong heat treatments all grains are prevailingly defined by (1 1 1) surfaces so that the functionalization of the film surface can be carried out in the same way as for the (1 1 1) surface.

8.1.5 The Surface of Porous Silicon

Porous silicon (π-Si) is a sponge-like system obtained by the dissolution of crystalline silicon upon anodization in a solution containing HF [221]. Its internal area can be 10^2–10^3 times its macroscopic geometrical area [222]. Pore diameter, porosity and porous-layer thickness depend sensitively on the electrolyte composition, bulk silicon doping (type and concentration), anodization current density, time, and temperature. Porous silicon is known to be hydrogen terminated, with spectroscopic evidence [223] for SiH_3, SiH_2. $(SiH)_2$ groups, thus being viewed as a mixture of hydrogenated 2×1 (1 0 0) Si and 1×1 (1 1 1) Si. Because of its room-temperature photoluminescence [128], π-Si attracted the interest of the scientific community as a light emitting material to be used in optoelectronic and photovoltaic devices [224]. However, due to its compatibility with existing silicon-processing technology [225], π-Si has also found extensive applications in several other fields, from biorecognition [226] to prosthetic systems [227] and chemical sensing [228]. Whatever the application is, the surface termination plays a crucial role: partial hydroxylation is known to completely suppress π-Si photoluminescence [229], while even mild chemical environments (e.g., aqueous solutions at physiological pH) can quickly etch away its surface. Grafted organic functionalities may impart specific tunable properties.

8.1.6 Inner Surfaces and the Fantastic Chemistry in Nanocavities

Even in an UHV,[4] surfaces can be preserved clean for limited duration: at a pressure of 10^{-10} Torr ($\simeq 10^{-8}$ Pa) each surface atom undergoes a collision with molecules of the residual atmosphere each at 10^4 s. Assuming a sticking coefficient of the order of unity, a freshly prepared surface can survive in such UHV for only a few hours. Although the prevailing gas in UHV is typically H_2 and this molecule has poor reactivity toward silicon, attaining and preserving clean silicon surfaces, and studying their evolution in the given atmosphere on long time scale, are, however, serious problems, whose solution seems impossible with ex situ probes.

This state of affairs has completely changed with the development of a process for the preparation of highly stable nanocavities with well-defined structures.

The implantation into solids of unreactive atoms with positive solution enthalpy and high diffusivity, like noble gases, followed by a thermal treatment at high temperature results first in an agglomeration of vacancies, followed by the formation of cavities and their filling with the implanted species. The implantation of noble gases into silicon produces highly pressurized cavities ("bubbles").

In the case of helium, the solution enthalpy and migration energy are sufficiently low to allow its almost complete out-diffusion from such bubbles [230, 231] after high-temperature treatment. What remains after this process are empty cavities whose shape, size, and distribution depend on the implantation conditions (energy and dose) and final heat treatment [232].

The evolution of the cavity shape during the thermal treatment is shown in Fig. 8.2. At the end of their evolution the cavities have mean diameter of a few tens of nanometers and attain the tetrakaidecahedron (truncated octahedron) equilibrium shape with the internal walls constituted mostly by (1 1 1), some (1 0 0), and few (1 1 0) planes [233].

In practice, nanometer-sized cavities with near-equilibrium shape may simply be prepared via a process involving the implantation of helium at high fluence (typically 1–4×10^{16} cm^{-2}) followed by heat treatment at high temperature (say $900°$C for 1 h) to allow the effusion of helium and the reorganization of the primeval bubbles to a shape close to what is expected from Wulff's construction. Figure 8.3 shows a picture of a cavity resulting from such a process. The figure shows a cavity with diameter of about 20 nm and shape defined by eight faces lying on (1 1 1) planes, two faces lying on (1 0 0) planes and rounded vertices.

In the considered example, the relative abundances of inner surfaces in relation to their orientations is 1 for (1 1 1), 1/4 for (1 0 0), and negligible for (1 1 0).

---

[4] "Vacuum" has several meanings. In technology it is not intended as the complete absence of matter; rather, in a vacuum system the concentration of matter is specified by an attribute: *rough vacuum* denotes an atmosphere at a pressure of 10^{-3}–1 Torr, *medium vacuum* corresponds to the pressure range 10^{-5}–10^{-3} Torr, *high vacuum* means a pressure ranging in the interval 10^{-8}–10^{-6} Torr, whereas pressures below 10^{-9} Torr define the *ultra-high vacuum* (see also footnote 10 in Chap. 3). The achievement of UHV requires, together with suitable pumping apparatuses, an accurate protocol for the control of the surface of the entire vacuum system.

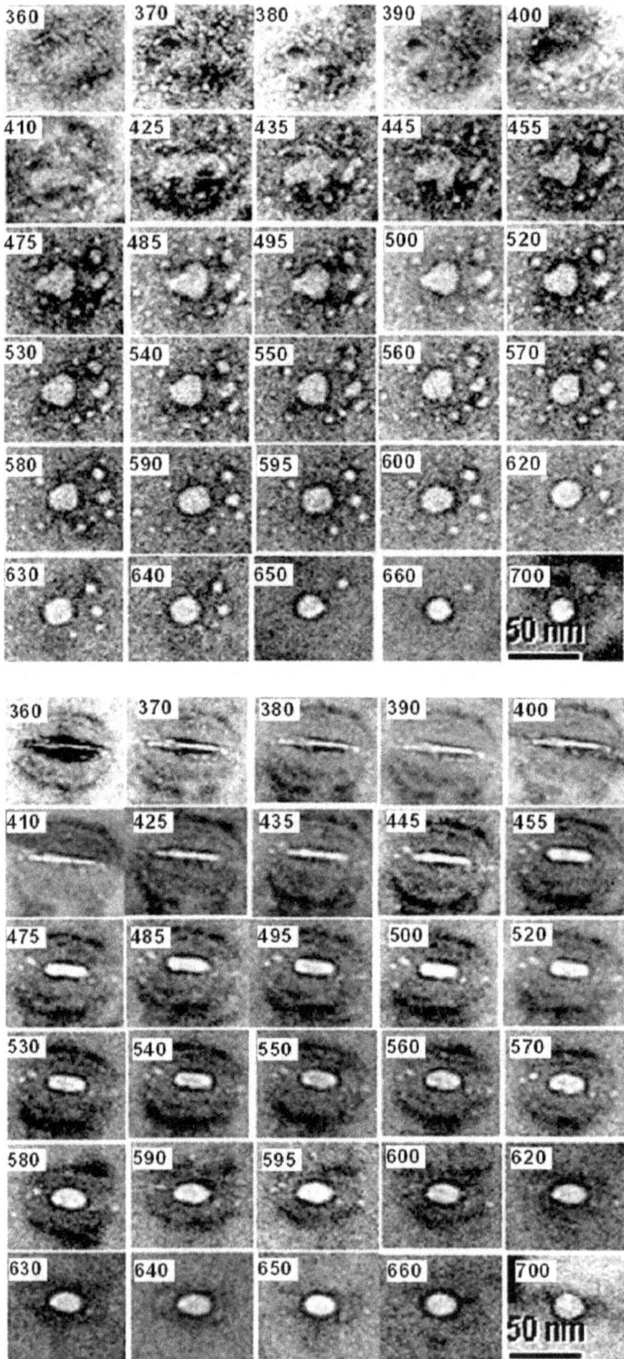

Fig. 8.2 Plan (*top*) and cross-section (*bottom*) views of a void resulting after consecutive thermal treatments at progressively higher temperatures (displayed in each picture) in (1 0 0) silicon implanted with helium at high fluence (after S. Frabboni et al., *Phys. Rev. B* **69**, 165209 (2004))

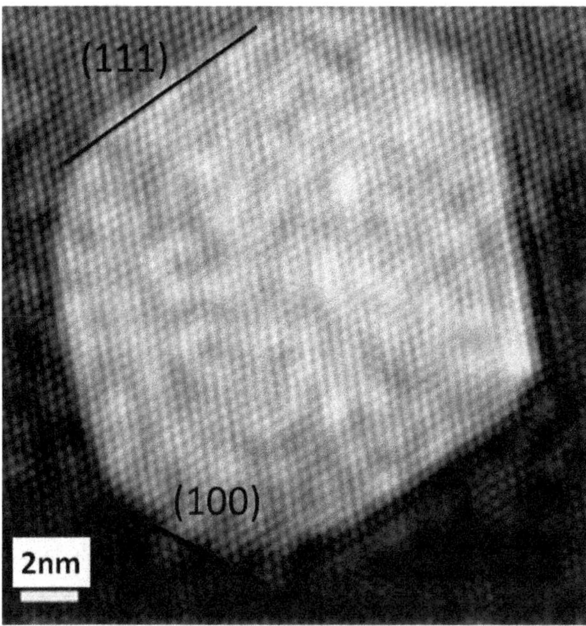

Fig. 8.3 Transmission electron micrography of a cavity produced by the helium implantation at $2 \times 10^{16}\,\mathrm{cm}^{-2}$ followed by heating at 900°C for 1 h (after S. Frabboni et al., *Phys. Rev. B* **69**, 165209 (2004))

Each inner surface may be contaminated by only the impurities originally contained in a small region surrounding it; assuming that all the oxygen atoms contained in a region of diameter larger than the one of the cavity by one order of magnitude has segregated in it, the inner surface should contain ten oxygen atoms at most. Clearly enough, this surface is protected against ambient contamination for a time exceedingly longer than the laboratory time scale.

Nanocavities formed inside a bulk material have thus ideal and stable surfaces, and are therefore an ideal medium for studying fundamental surface processes, like reconstruction, passivation and reactions with species eventually introduced in the cavities. Over (conventional) outer surfaces, the inner surfaces of cavities in solids have the advantage of not being exposed to environmental contamination; they are therefore almost ideal vehicles for the study of superficial properties in ambient situation not achievable even under UHV conditions. A paradigmatic example is provided by the use of such determination of the surface tension of low Miller-index surfaces of silicon [234–236].

The obvious difficulty comes from the fact that such surfaces can hardly undergo chemical treatments and can only be sensed by probes that are not absorbed by the embedding solid. Among the very few methods for the chemical treatment of such surfaces, ion implantation plays a special role. In fact, the physical and chemical properties of the inner surfaces may be controlled by atomic species implanted in the silicon at an energy so chosen as to allow them to stop in the region of cavities and segregating preferentially inside the cavity or at its surface after the thermal treatment required for the annealing of the radiation damage. Hydrogen, fluorine, and nitrogen in silicon seem to satisfy these conditions.

Hydrogen Hydrogen is especially interesting because of the following facts:

- The production yield of defects imparted to silicon by the implantation of hydrogen is minimum over the Periodic Table, so that it is expected to give the minimum of interference on segregation phenomena.
- The hydrogen termination of silicon surfaces continues to be a hot research topic [199, 237, 238] that can hardly be studied in the temperature and pressure conditions that, instead, can be easily achieved in cavities.
- The Si–H stretching modes lie in a spectral region, centered on about $2000\,\mathrm{cm}^{-1}$, where the absorption of silicon is so small to allow a spectroscopic investigation via infrared absorption.

The implantation of hydrogen in silicon produces a complex pattern of hydrogen configurations, bonded to otherwise perfect silicon (bond centered, H_{BC}; antibonding, H_{AB}; H_2; or H_2^* – a close H_{BC}–H_{AB} pair) or to implantation defects (vacancy or self-interstitial clusters) [239].

Thermal treatments at moderate temperature (say up to 600°C) produce the agglomeration of vacancy clusters to form cavities eventually filled (or whose formation is sustained) by the H_2 formed by the decomposition of defect-hydrogen centers [240]. This process might result in the formation of pressurized regions (with pressure of 10^2–10^3 atm) at temperature up to 600°C otherwise unattainable in static conditions. In these conditions hydrogen can interact with silicon at the walls of the cavities inducing the formation of chemical bonds.

That hydrogen interacts really with cavities is demonstrated by infrared investigation of Romano et al., showing the appearance of the signatures attributed to highly ordered hydrogen-terminated 2×1 (1 0 0) and 7×7 (1 1 1) surfaces [240] (Fig. 8.4).

Fluorine The implantation of fluorine at low fluence is known to produce a fluorine-rich layer from which the effusion of fluorine is quite easy [241, 242]. The implantation of fluorine at a depth just below the cavity region followed by a heat treatment at a temperature high enough to allow its diffusion will thus produce the segregation of fluorine inside the cavities, empty or filled with H_2. It will produce molecular fluorine and F–H compounds, strongly stabilized by hydrogen bonds and with stretching accessible to infrared spectroscopy. The F–H compounds can interact with silicon atoms at the inner surface of the cavities thus allowing a study of the complex Si–F–H chemistry. Moreover, the preparation is carried out in a way that renders possible the study of the hydrogen bond (for the species that manifests it at the maximum degree) even in extreme ambient conditions.

The cavities can be formed at temperatures, around 1,000°C, at which the segregation of the interstitial oxygen contained in the silicon wafer is possible at the inner surface of the cavities. The possible formation of Si–F–O–H compounds opens new interesting research opportunity.

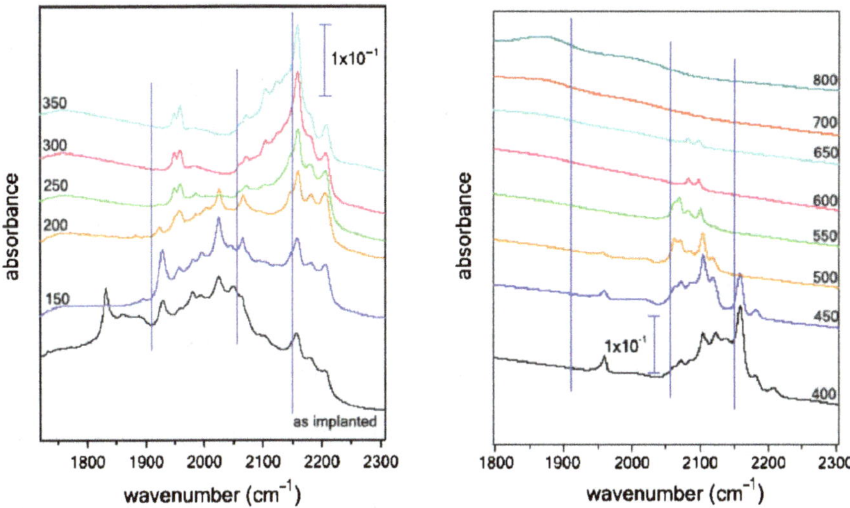

Fig. 8.4 Infrared spectra from silicon targets containing a nanocavity array postimplanted with hydrogen after subsequent isothermal treatments each lasting 1 h (by courtesy of E. Romano)

Nitrogen More exciting is the chemistry allowed by nitrogen implantation. In fact, it is known that nitrogen implanted at low fluence is a fast diffuser even at moderate temperatures, that makes its segregation at inner surfaces possible [243, 244]. A pre- or post-implantation of hydrogen thus makes possible the infrared investigation of the expected complexes (e.g., Si–N, Si–N$^\bullet$H, Si–NH$_2$, Si–N(H)–Si, etc.) or molecules (N$_2$, HN=NH, H$_2$N–NH$_2$, NH$_3$, etc.) that are formed after or during suitable subsequent thermal processes.

Neutral Ultra-Hot Matter Even more exciting is the investigation of the chemical phenomena occurring in the pressurized cavities after the passage of heavy ions therein. Heavy molecular ions in a stopping regime dominated by nuclear collisions produce indeed columns of neutral ultra-hot matter (with temperature up to 10^5 K) lasting 10^{-11} s for which a new chemistry is expected to occur [245].

8.2 Routes for Silicon Functionalization

Although some elementary form of silicon–organic chemistry can be studied even in cavities,[5] the richness of reaction allowed by the available organic molecules is accessible only at the external surfaces.

Several methods have been proposed for the functionalization of silicon; in the following the attention is focused on those leading to direct Si–C covalent bond, instead of oxygen-mediated Si–O–C bond.

---

[5] For instance, by the implantation of molecular ions of alkanes into silicon containing pressurized cavities at an energy allowing the stopping in the bubble region.

Some of them (like those involving the exposure of halogen-terminated silicon to Grignard reagents or organolithium compounds) are manifestly incompatible with the underlying CMOS structure because of waste products (magnesium or lithium halides). Among the others, the following ones are considered:

The **cycloaddition** of molecules with alkene or alkyne functions to the clean, 2×1 reconstructed, (1 0 0) oriented, silicon surfaces [246]

The **condensation** of molecules containing alcoholic groups at hydroxyl-terminated silicon surfaces [247]

The **silanization** of

- Hydroxyl-terminated surfaces [248] as well as of hydrogen-terminated surfaces [249] via reaction with chlorosilane-terminated molecules and elimination of HCl
- Hydrogen-terminated surfaces via reaction with azanes and elimination of NH_3 [249]

The **arylation** of hydrogen-terminated silicon via its reaction with diazonium salts and elimination of HBF_4 and N_2 [105]

The **hydrosilation** of alkenes at hydrogen-terminated surfaces [250].

Figure 8.5 sketches all these routes.

If compatibility with poly-Si is a major issue, cycloaddition is immediately discarded from these routes. In fact, cycloaddition is limited to the extremely reactive, *clean, 2×1 reconstructed, (1 0 0) surface of single crystalline silicon*, which can hardly be imagined to be producible at the poly-Si surfaces. (Actually, the adoption of the method described in [123] for silicon-on-insulator substrates would allow the use of cycloaddition for molecular grafting on the lower silicon-wire array.)

As far as no process is known for the controlled preparation of *hydroxyl-terminated silicon surfaces*, condensation on and silanization of these surfaces seem difficult to implement too. Discarding these routes is also supported by the following reasons: the Si–O bond allows water insertion (the preliminary step of silicon oxide formation whose insulating nature would deteriorate device performances); the Si–O–C bridge undergoes hydrolysis in the presence of water; and the high polarity of the Si–O bond is the responsible for an increase of the energy barrier to electron drift [105].

On the contrary, as discussed in the previous section, device quality, relatively stable, *hydrogen-terminated silicon surfaces* may be prepared by etching with aqueous solution of HF, HF_{aq}, a sacrificial film of SiO_2 thermally grown on the surface [208]. This suggests the use of such terminations for the functionalization.[6]

Of the considered processes, the ones involving the diazonium salts and chlorosilanes seem difficult to implement. Integration issues, indeed, locate the

---

[6] Whereas quality and stability are generally high for the (1 1 1) surface etched with buffered HF_{aq} [207], they are poor for the (1 0 0) surface [173]. The quality of these surfaces can however be improved by exposing them to H_2 at moderate (700–800°C) [204] or high (850–1,100°C) [238] temperatures.

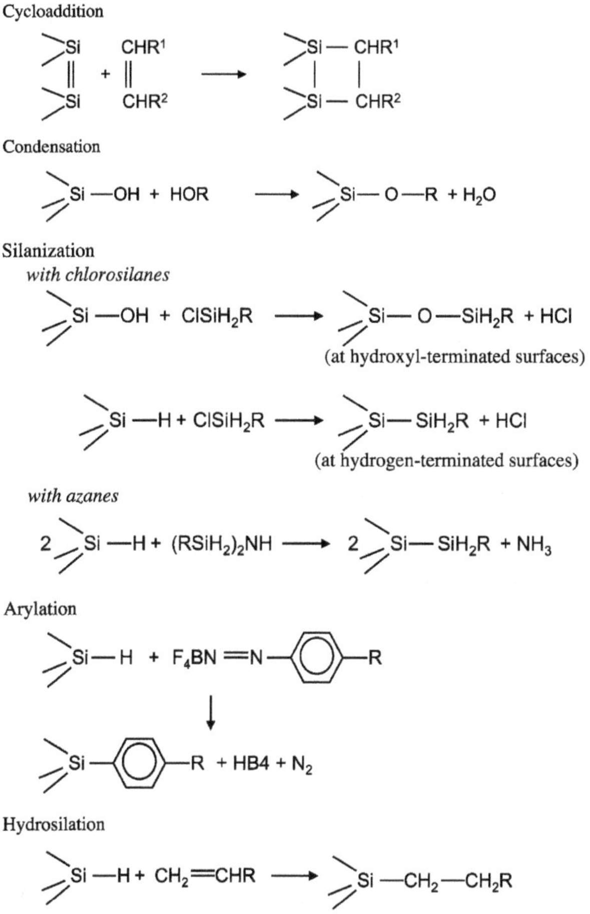

Cycloaddition

Condensation

Silanization
with chlorosilanes

(at hydroxyl-terminated surfaces)

(at hydrogen-terminated surfaces)

with azanes

Arylation

Hydrosilation

Fig. 8.5 The considered pathways for silicon functionalization

derivatization in the back-end stage of the process [109], so that the waste reaction products (HCl and BF_3) could react, even vivaciously, with the circuit metallizations.

Commercial availability of precursors of functional molecules (larger for alkene-terminated molecules than for azanes) thus makes *hydrosilation of unsaturated hydrocarbons at the hydrogen-terminated silicon surface* the most serious candidate process for silicon functionalization.

8.2.1 Hydrosilation

Hydrosilation is not limited to alkenes but occurs in general for molecules involving C=C, C≡C, C=O, C=S, or C=N groups [251]. Limiting the attention to alkenes and alkynes, this process may mechanistically be divided into three steps: a π bond of a C=C or C≡C group in the vicinity of a silanic group is cleaved; silanic hydrogen shifts from silicon to one of the carbon atoms originally involved in the π bond; and the unsaturated silicon and carbon atoms bind to one another. Actually, in the homogeneous phase the regiochemistry of addition is reported to be *anti*-Markovnikov, i.e., silicon goes on the carbon with the most hydrogen atoms [182, p. 402].

While in homogeneous phase internal alkenes undergo hydrosilation (although at a lower rate than for terminal alkenes [182, p. 402]), at the surface of hydrogen-terminated silicon alkene hydrosilation is practically impossible unless the π bond is terminal, due to steric constraints. Heterogeneous hydrosilation is carried out with 1-alkenes or 1-alkynes; the use of other monomers would indeed be less convenient because of the resulting Si–O, Si–N, or Si–S surface dipoles.

Even though for the hypothesized application one should be mainly involved in the hydrosilation at hydrogen-terminated poly-Si, the process is better understood at the surface of single-crystalline silicon. Most studies have been carried out functionalizing the hydrogen-terminated (1 1 1) or (1 0 0) silicon surfaces with 1-alkenes [252–261] or 1-alkynes [262–269].

Of the various terminations undergoing hydrosilation, the alkyne ones offer a great advantage: if they are the tails of π-conjugated molecules (and if they are kept normally oriented with respect to the surface), the terminal π conjugation is not destroyed even after the first cleavage of the π-bond; moreover, if the surface has a prevailing (1 1 1) orientation, there is no other Si–H group so close to the grafted molecule to allow the cleavage of the residual π bond. Because poly-Si has usually a prevailing (1 1 1) orientation, *the optimal combination for molecular electronics is poly-Si in concert with alkyne-terminated π-conjugated molecules.*[7]

8.2.2 Hydrosilation at the Hydrogen-Terminated (1 0 0) Si Surface

Even though neither the hydrosilation mechanism nor its dependence on surface preparation is fully understood yet [250, 270, 271], there are already indications about bond distribution and stability of the derivatized surfaces. The information given in the following applies to the behavior of the derivatized (1 0 0) Si surface; however, a similar, or even better, behavior is also expected not only for the (1 1 1) Si surface (whose higher environmental stability with respect to (1 0 0) Si is generally recognized) but also for the surface of poly Si, because it is (1 1 1) oriented. It is also noted that an enormous number of molecules have been grafted to silicon; the attention in the following is concentrated on simple species, because they contain all the information contained in larger species, and are easier to interpret.

Figure 8.6 compares the X-ray photoelectron spectroscopy (XPS) Si 2p spectra from the following samples: (a) a pristine hydrogen-terminated (1 0 0) Si surface;

---

[7] It is, however, noted that even the use of aryl diazonium salts, as in [105], does not interrupt the π conjugation along the molecular chain.

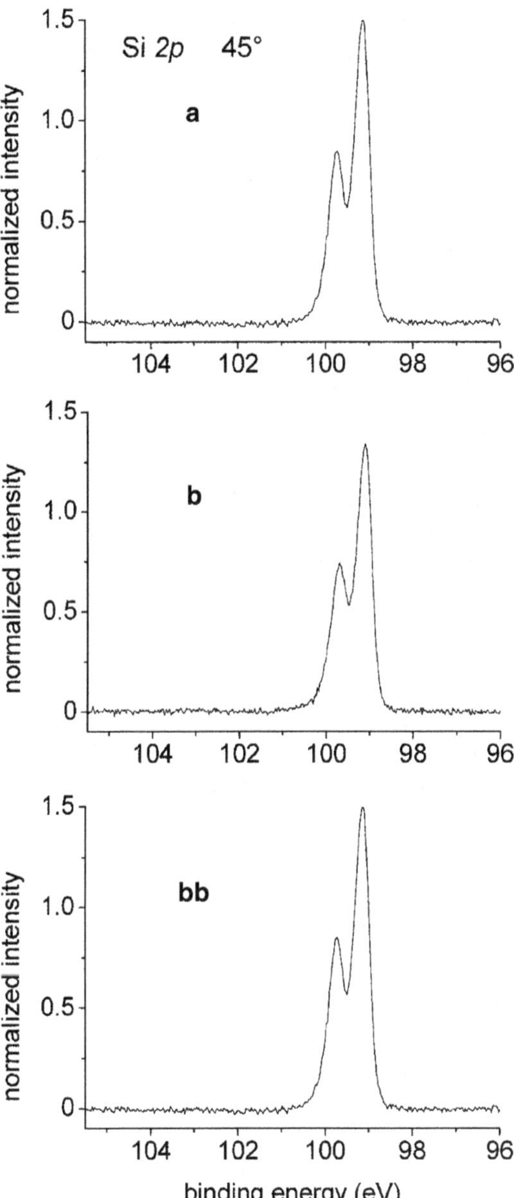

Fig. 8.6 Si 2p XPS spectra from: (**a**) a hydrogen-terminated (1 0 0) silicon surface; (**b**) the same surface after derivatization with 1-octene; and (**bb**) the hydrogen-terminated surface after derivatization with 1-octyne. All spectra are normalized to the Si 2p total intensity (after G.F. Cerofolini, E. Romano, *Appl. Phys. A* **91**, 181 (2008))

(b) the hydrogen-terminated surface after functionalization with 1-octene; and (bb) the hydrogen-terminated surface after functionalization with 1-octyne. In both cases the functionalization was achieved via the immersion of the hydrogen-terminated surface in the pure hydrocarbon kept at 170°C for 2 h.

Figure 8.7 shows the corresponding C 1s XPS spectra. The quantitative analysis of these spectra (as well as of the entire family of angle-resolved spectra) is given in [269].

In the following the attention will be focused on a few observations which prove the possibility to derivatize silicon and the long-term environmental stability of the derivatized surfaces.

Figure 8.8 shows the C 1s XPS spectra from a hydrogen-terminated (1 0 0) Si surface and from the same surface after derivatization with 1-octyne (the spectrum from the pristine surface is due to adventitious carbon physically or chemically adsorbed from the atmosphere). The figure also shows the effect of aging: as the exposure of the pristine surface to the environment results in a progressive capture of hydrocarbons containing carbon in C–O, C=O, and O=C–O configurations (with C 1s binding energy centered on ca. 286.8, 288.3, and 289.8 eV, respectively, the binding energy of alkanic carbon being centered on 285.0 eV), the C 1s spectrum of the grafted alkyne remains remarkably stable, showing only a minor increase of the oxidized C–O fraction.

Figure 8.9 shows the Si 2p XPS spectra from the samples whose C 1s spectra are shown in Fig. 8.8. The figure shows that grafting improves the environmental stability of silicon too: while the prolonged exposure of the pristine surface to the environment results in the formation of an oxide film of a few monolayers (with the appearance of a Si 2p signal centered around 103 eV), the same exposure of the grafted surface results in a film with lower thickness by a factor of about 3.5.

8.3 Grafting in Restricted Geometries

The idea of inserting the functional molecules between predefined contacts was first formulated by Cerofolini and Ferla with the aim of exploiting nanometer-sized features [106] commonly obtained in CMOS processing; experimental evidence for the feasibility of this process was given in [272]. Transforming a laboratory practice into a technology, however, is not trivial.

Both the deposition of a sacrificial slab with thickness matched to the functional molecules and the grafting in recessed regions with typical dimension on the nanometer length scale seem indeed difficult tasks. The first problem, however, is easily overcome by:

- Preparing the crossbar via the XB* process
- Using as sacrificial vertical spacer a thermally grown SiO_2 slab (as demonstrated in [273], the growth of oxide films at the silicon surface can be controlled with monolayer accuracy)
- Etching it at the end of the process with aqueous solution of HF (HF_{aq})
- Rinsing in H_2O

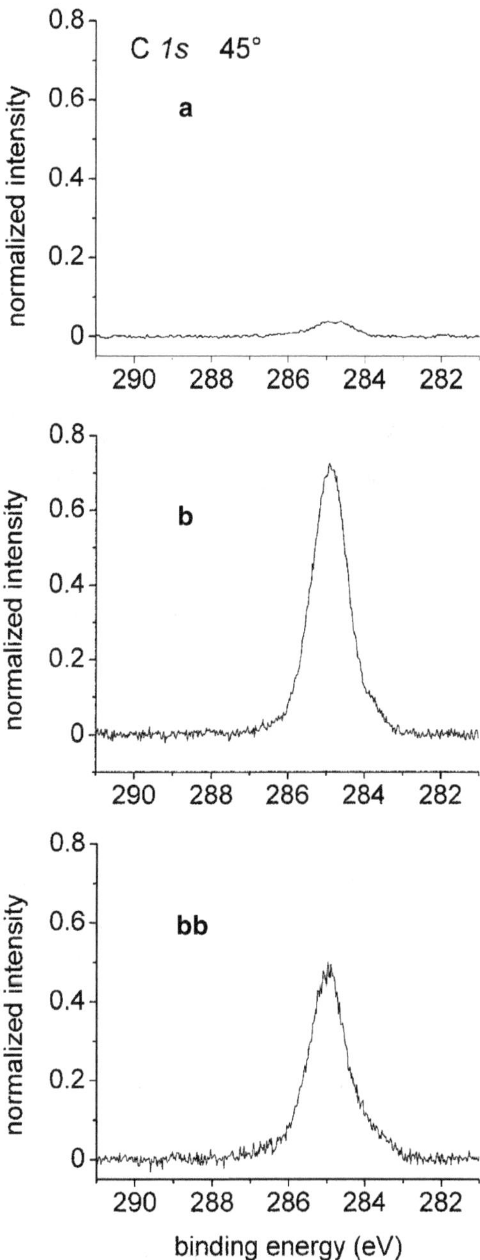

Fig. 8.7 C 1s XPS spectra from: (**a**) a hydrogen-terminated (1 0 0) silicon surface; (**b**) the same surface after derivatization with 1-octene; and (**bb**) the hydrogen-terminated surface after derivatization with 1-octyne. All spectra are normalized to the Si 2p total intensity (after G.F. Cerofolini, E. Romano, *Appl. Phys. A* **91**, 181 (2008))

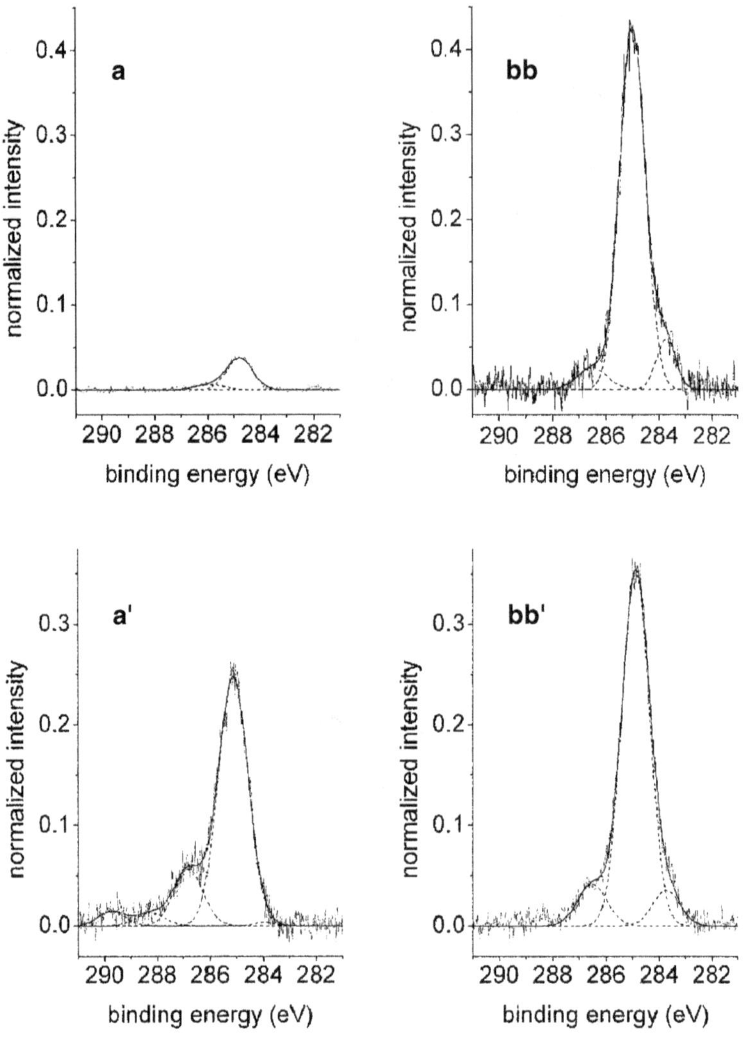

Fig. 8.8 C 1s XPS spectra from: (**a**) a hydrogen-terminated (1 0 0) silicon surface; (**bb**) the same surface after derivatization with 1-octyne; (**a′**) a surface as (**a**) after prolonged (2.6×10^7 s) to air; and (**bb′**), a surface as (**bb**) after prolonged (2.6×10^7 s) to air. All spectra are normalized to the Si 2p total intensity (after G.F. Cerofolini, E. Romano, *Appl. Phys. A* **91**, 181 (2008))

Figure 8.10a sketches the cross section of the active zone before HF_{aq} attack, whereas Fig. 8.10b shows the same zone emphasizing the Si–H and Si–OH terminations of the opposing poly-Si surfaces resulting after rinsing.

It is noted that the etching with HF_{aq} results in prevailing Si–H, hydrophobic, terminations. Because of this situation, if the distance between hydrogen terminations on opposing sides is comparable or smaller than the diameter of the solvated species

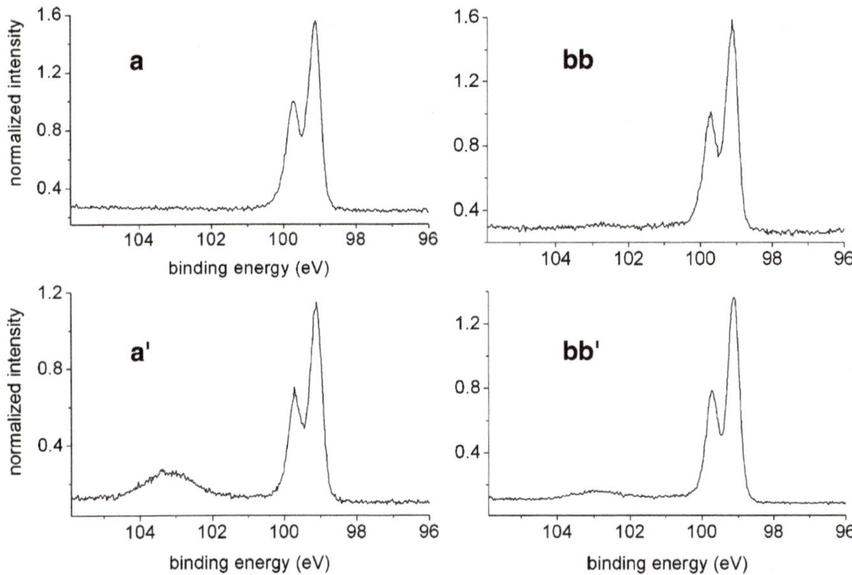

Fig. 8.9 Si 2p XPS spectra from: (**a**) a hydrogen-terminated (1 0 0) silicon surface; (**bb**) the same surface after derivatization with 1-octyne; (**a′**) a surface as (**a**) after prolonged (2.6×10^7 s) to air; and (**bb′**), a surface as (**bb**) after prolonged (2.6×10^7 s) to air (after G.F. Cerofolini, E. Romano, *Appl. Phys. A* **91**, 181 (2008))

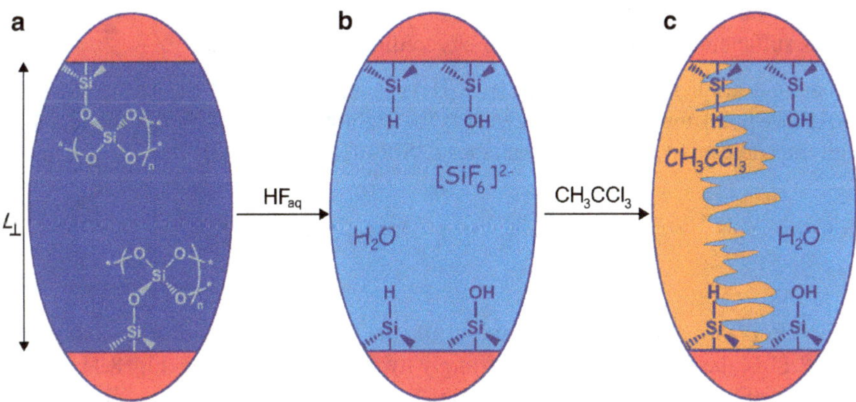

Fig. 8.10 Cross section of a crosspoint before HF$_{aq}$ attack (**a**), immediately after attack (**b**), and during the substitution of the aqueous phase with polar aprotic solution (**c**)

responsible for the etching [213, 215] the reaction is expected to be kinetically locked. This hindrance may however be removed by the addition of appropriate surfactants.

The attachment of the reconfigurable molecules to poly-Si walls separated by a distance $L_\perp$ of a few nanometers suffers from the risk of being a mere declaration of will until and unless a process for that is sketched, if not specified in detail.

Assume that the molecule (or an appropriate precursor) is characterized by a polar core and two hydrophobic terminations (this situation is characteristic of the molecules considered in the previous parts: the ones based on the bipyridinium core, irrespective of their alkyl or π-conjugated tails, and that with nitroaniline core). This structure can be exploited to facilitate the segregation and appropriate orientation of the functional molecules inside the cavities defined by the silicon walls.

Water is expected to interact strongly with polar or ionic sites of the reconfigurable molecules, so that its elimination seems mandatory. This process, however, is not so simple as it appears at the first glance: due to the interfacial tension $\gamma_{s|w}$ at the solid–water interface, menisci with curvature radius of R the order of $L_\perp$ are expected to form: $R \approx L_\perp$. For typical values of $\gamma_{s|w}$ and $L_\perp$ ($\gamma_{s|w} = 10^{-3}$–$10^{-1}\,\mathrm{J\,m^{-2}}$ and $L_\perp \approx 3\,\mathrm{nm}$), the pressure P resulting from the Laplace equation,

$$P = 2\gamma_{s|w}/R,$$

may therefore be so high, $P = 10^7$–$10^9\,\mathrm{Pa}$ (10^2–10^4 atm), as to destroy the structure. Even though these forces might in principle be canceled by eliminating the water via supercritical extraction or with the addition of suitable surfactants that make $\gamma_{s|w}$ negligible, another route is hypothesized. This route does not require the removal of the trapped water, but rather its controlled replacement.

The first step consists in the substitution of another liquid ℓ for water (Fig. 8.10c) satisfying the following conditions:

- ℓ does not react with the silicon surface.
- ℓ does not react with the chlorosilanes.
- ℓ is sufficiently polar to dissolve the reconfigurable molecules.

Tentative examples for ℓ are CH_2Cl_2, $(CH_3)_2CCl_2$, etc.

At this point the silicon surfaces may be stabilized by passivating the silanols with the addition of trialkylchlorosilanes $ClSiR_3$ ($R = CH_3$, C_2H_5, etc.):

$$\overset{\textstyle\backslash}{\underset{\textstyle/}{-}}Si-OH + ClSiR_3 \longrightarrow \overset{\textstyle\backslash}{\underset{\textstyle/}{-}}Si-O-SiR_3 + HCl.$$

Thus after the complete substitution of the solvent for the water solution (Fig. 8.11d), the surfaces acquire a strongly hydrophobic character (Fig. 8.11e).

After HF_{aq} attack, rinsing, substitution of ℓ for water, and surface passivation with trialkylsilicon termination, the structure of the recessed region is sketched in Fig. 8.12.

This situation is similar to that of the molecule (hydrophobic tail and polar core), so that the total free energy of the system containing the molecule is expected to decrease with the substitution of the functional molecule for the solvent (Fig. 8.11f). Moreover, if this process is carried out with stoichiometric amounts of hydrogen terminations and molecules, the latter are expected to segregate almost uniquely in the recessed region rather than on other hydrophobic sites (because of the free energy excess of the other hydrophobic tail of the molecule adsorbed on the vertical walls

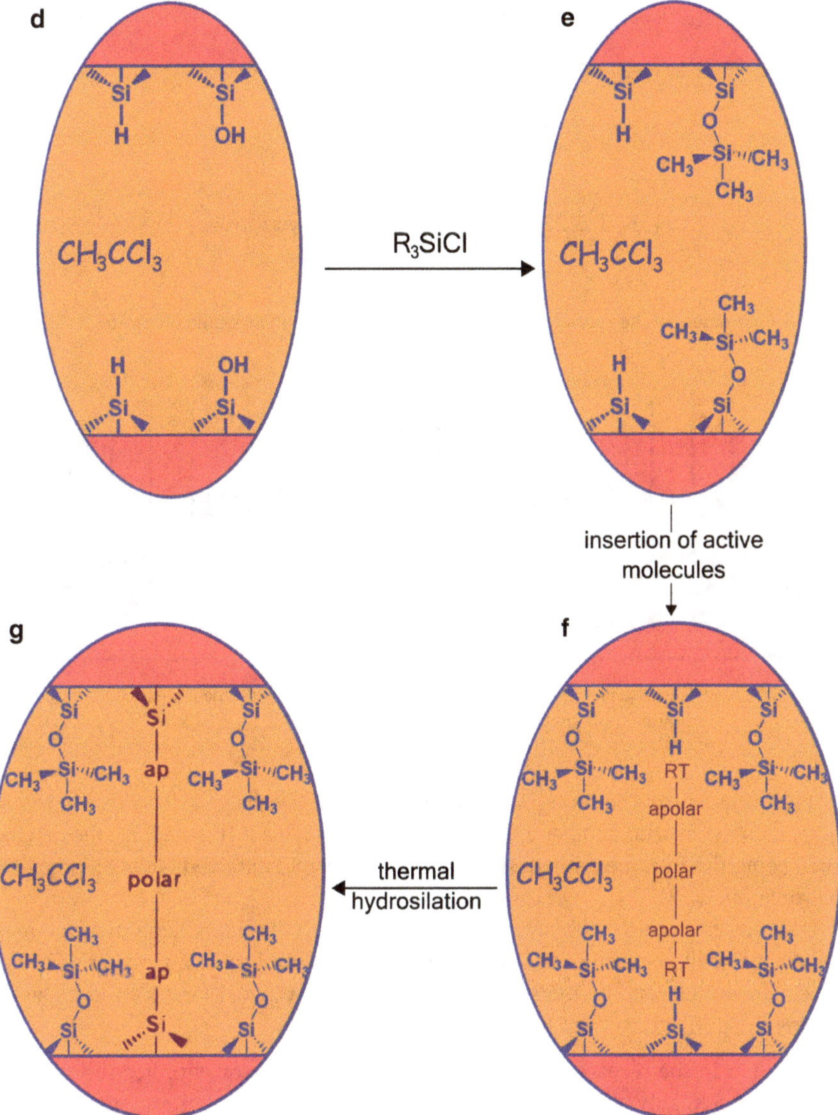

Fig. 8.11 The process for the segregation and self-assembly of the functional molecules into crosspoints [RT in **f** denotes a reacting termination (alkene, alkynes,...) undergoing hydrosilation]

of the spacers). It is noted that surface tension works to eliminate the reconfigurable molecules from the regions separating the spacers only if the lateral wall separation is higher than twice the molecular length l_{mol}. This provides a crossbar design rule:

$$t_{sp} - t_{Si} > 2l_{mol},$$

the symbols having the same meaning as in Sect. 6.2.

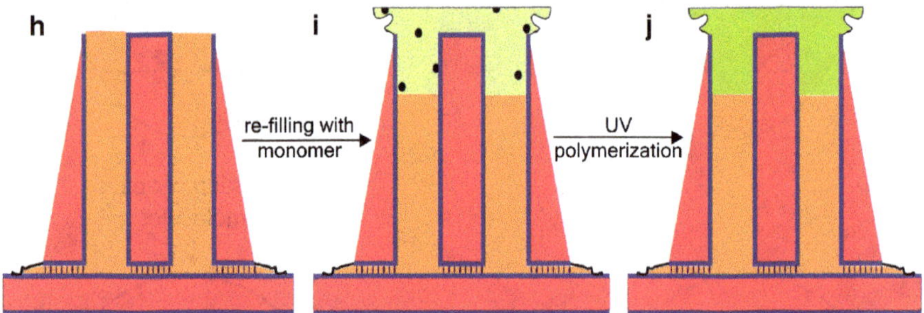

Fig. 8.12 Structure of the recessed region and its hydrophobic–hydrophilic properties

Fig. 8.13 Completing the process of crosspoint functionalization with the formation of a solid cap encapsulating the liquid embedding the active molecules

This process may be brought to completion with the progressive substitution of the polar solvent with another completely nonpolar solvent (Fig. 8.13h), the driving force being the free energy excess of the polar solvent contained between nonpolar hydrophobic walls.

Inserting molecules with size perfectly adapted to the intersilicon distance may be very difficult because of the steric constraints posed by the nonplanarity of the opposing surfaces and by their alkyl terminations. The underlying idea to make up for this difficulty consists of the following steps:

- Insertion in the recessed region of a molecular precursor with head-to-tail distance appreciably shorter than the intersilicon separation, but roughly the same as the distance between their alkyl terminations
- Transformation of the precursor into the reconfigurable molecule via physical (e.g., thermal) or chemical (e.g., pH) stimulation

An example of the way that can be achieved is discussed in Sect. 9.2.

After a heat treatment at moderate temperature (say, $< 100°$C) to complete the segregation, the process ends with the hydrosilation of both tails of the molecule at silane termination, as shown in Fig. 8.13h (this process is specified better in the example of Sect. 9.2). The whole structure is automatically passivated if the last nonpolar solvent undergoes polymerization during the final heat treatment (Fig. 8.13i). Figure 8.13 sketches the final stages of this process.

Of course, there are some degrees of freedom in the order of hydrosilation and polymerization. It is possible that, whichever monomer or process is used for the polymerization, the process destroys the functional molecules. A more conservative process might therefore be the following: after controlled evaporation of a small amount of nonpolar solvent from the open channels, they are refilled with a not miscible polar solvent; saturating this solvent with $SiCl_4$, adding water, and curing at moderate temperature (say, 150–200°C) will produce the formation of a $SiO_{2-x}(OH)_{2x}$ crust able to embed the underlying organic mixture.

Although the entire process is designed not to manifest capillary effects, the crossbar structure, formed by an array of suspended beams (of length in the micrometer length scale) separated by the underlying beams by a few nanometers, is manifestly delicate and suffers from the risk of collapsing.

Countermeasures to avoid the collapse of the structure can however be adopted. For instance, the bottom array can be formed by a sequence of poly-Si wires separated by insulators formed by a material (like pyrolitic Si_3N_4 or Al_2O_3) not undergoing oxidation and etched when etching the sacrificial SiO_2 film separating bottom and top arrays. In so doing they behave as pillars sustaining the upper wires.

8.4 Three-Terminal Molecules

The use of molecules in molecular electronics is essentially due to the fact that they embody in themselves the electrical characteristics of existing devices. Characteristics of nonlinear resistors, diodes, and Schmitt triggers have been reported for two-terminal molecules; their use as nonvolatile memory cells is possible thanks to the stabilization of a metastable state excited by the application of a high voltage (thus behaving as a kind of virtual third terminal). Three-terminal molecules would offer more application perspectives non only because they could mimic transistors but also manifest genuine quantum phenomena like the Aharonov–Bohm effect.

The application potentials of three-terminal molecules can however be really exploited only if all terminals can be contacted singularly. This is manifestly impossible using the XB or XB_+ routes, but is possible in the XB_* framework. The major advantage of poly-Si in the XB_* route is that with this material there is no problem of metal electromigration. It will however be shown in the following that the multispacer technology can be adapted for the preparation of nanowire arrays of poly-Si and of metals in an arrangement that

- Avoids the problem of metal electromigration
- Facilitates the self-assembly of functional molecules
- Allows the use of three-terminal molecules

Assume that the top array defining the crossbar is formed by poly-Si nanowires whereas the bottom array has a more complicated structure: in each SPT cycle the conformal deposition is formed by the following multilayer:

$$SiO_2 \mid poly\text{-}Si \mid SiO_2 \mid metal \mid SiO_2 \mid poly\text{-}Si.$$

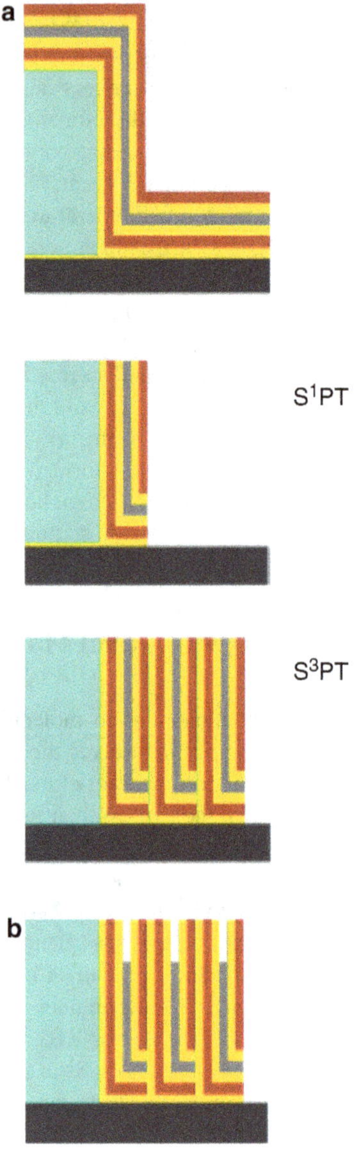

Fig. 8.14 Multinanoarray formed by application of the SPT to multilayered (insulator, poly-silicon, insulator, metal, insulator, poly-silicon) films. After completion of S^nPT (**a**) a selective etch is used to create a shallow trench in the metal (**b**)

The metal might be, for instance, platinum deposited via decomposition of the PtF_6 precursor.

After three SPTs, one gets the structure sketched in Fig. 8.14a; a subsequent time-controlled selective etch of the metal electrode will result in the formation of a recessed region, as sketched in Fig. 8.14b.

Observe now that the thiol-terminated molecules self-assemble spontaneously on many metals (like platinum or gold, etc.) forming regular monolayers. In this way if the considered three-terminal molecules are formed by two alkyne and one

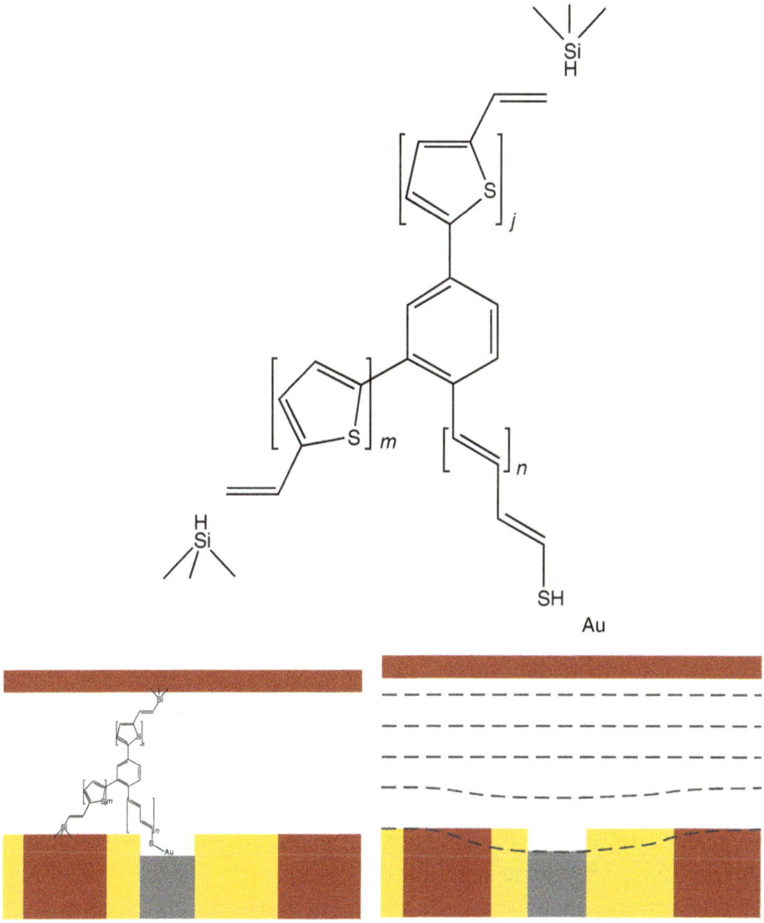

Fig. 8.15 Three-terminal molecular device formed by reaction of a suitably terminated molecular switch (*top*) with hydrogen-terminated silicon and gold electrodes (*bottom and left*) so shaped as to reduce the electric field at the metal electrode (*bottom and right*)

thiol terminations, they can arrange themselves in the crosspoints in an ordered way allowing their covalent grafting by simple heat treatment, as sketched in Fig. 8.15. This figure also shows how the electric field at the metal–sulfur interface may be significantly lower than at the silicon–carbon interface.

8.5 Nanostructured Oxo-Bonded Silicon

As already mentioned at the beginning of this chapter, silicon and carbon share more differences than analogies. Rather, the analogies are stronger between oxo-bonded silicon and carbon, provided that the oxo bridge between silicon atoms is considered

Table 8.2 Formal analogy between hydrocarbons and siloxo compounds

Name	Formula	Name	Formula
Alkane:		*Alksoxan:*	
Methane	CH_4	Methsoxan	$Si(OH)_4$
Ethane	CH_3CH_3	Ethsoxan	$Si(OH)_3OSi(OH)_3$
Propane	$CH_3CH_2CH_3$	Propsoxan	$Si(OH)_3OSi(OH)_2OSi(OH)_3$
Polyethylene	$(CH_2CH_2-)_n$	Polyethsoxylen	$(Si(OH)_2OSi(OH)_2O-)_n$
Polypropylene	$(CH(CH_3)CH_2-)_n$	Polypropsoxylen	$(Si(OH)(OSi(OH)_3)OSi(OH)_2O-)_n$
Alkene:		*Alksoxen:*	
Ethene	CH_2CH_2	Ethsoxen	$Si(OH)_2O_2Si(OH)_2$
Propene	CH_2CHCH_3	Propsoxen	$Si(OH)_2O_2Si(OH)OSi(OH)_3$
Polyacetylene	$(CH=CH-)_n$	Polyacetsoxyle	$(Si(OH)O_2Si(OH)O-)_n$
Cumulene	$(C=C=)_n$	Cumulsoxen	$(SiO_2Si(O-)_2)_n$
Alkyne:		*Alksoxyn:*	
Ethyne	$CHCH$	Acetsoxyn	$Si(OH)O_3Si(OH)$
Propyne	$CHCCH_3$	Propsoxyn	$Si(OH)O_3SiOSi(OH)_3$
Polyyne	$(C{\equiv}C-)_n$	Polysoxyn	$(SiO_3SiO-)_n$

as the equivalent of a bond between carbon atoms. This analogy is made possible by the near independence of the oxo bridge energy on the $\widehat{Si-O-Si}$ angle. From the point of view of the ability to catenation, the silicon species more similar to alkanes are siloxanes, where silicon atoms are linked through oxo bridges $-O-$. In view of the weak dependence of the energy content of the oxo bridge on angle $\widehat{Si-O-Si}$, various allotropic forms of silica and siloxo compounds are indeed possible. It is immediately seen that alkane, alkene and alkyne species have counterparts in siloxo compounds, for which the names alksoxan, alksoxen, and alksoxyn seem the most natural. In this case the silicon analogues of cumulenes and polyyne (organic compounds with alternating single and triple bonds) are linear or cyclic $(SiO_2SiO_2)_n$ or $(SiO_3SiO)_n$, which seem to be (or already are) in the reach of the siloxan chemistry even for $n > 3$. The major difference is the absence of aromaticity in siloxo compounds. The analogy is displayed in Table 8.2. An even more intriguing analogy is discussed in [274].

8.5.1 Hydrothermal Synthesis: Zeolites

The energetic properties of the $\widehat{Si-O-Si}$ angle allow the formation of a huge variety of nanostructures via relatively simple hydrothermal methods using templating molecules which impart to the embedding silica their complementary shape. Moreover, the hydrothermal method allows the doping of the silica structure with other oxides; oxides of Group III (e.g., B_2O_3, Al_2O_3) leave necessarily on the silica silanol defects with acidic (often very acidic) character (see Fig. 8.16), whereas oxides of Group V (e.g., P_2O_5, As_2O_5) result necessarily in OH terminations with basic character.

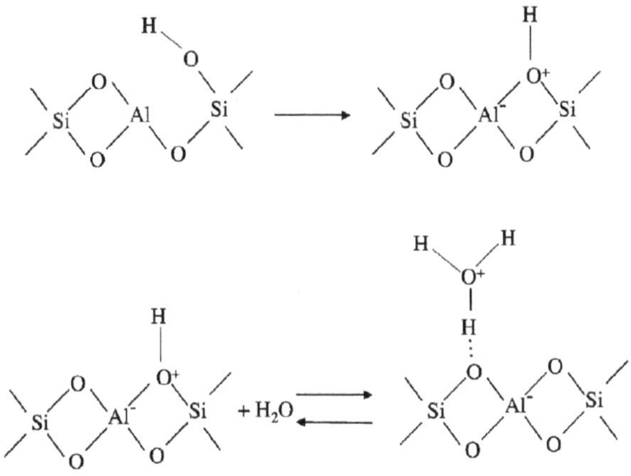

Fig. 8.16 Formation of a Brønsted acid from the aluminum Lewis acid (*top*) and the subsequent dissociation via reaction with a water molecule (*bottom*)

When the resulting solid is crystalline (that happens quite frequently, with the templated voids with size on the nanometer length scale), the structure is usually referred to as a *zeolite*.

Although at this moment no attempt at integrating the zeolite and IC technologies is known, from the point of view of their constituting materials they appear highly compatible, and the integration seems potentially useful for hybrid bio-ICs. In fact, biological systems communicate via exchange of protons, rather than electrons, and not only are protons fast diffusers in oxides but they are also easily available in acidic zeolites.[8]

8.5.2 Hydrolysis and Polycondensation: Aerogels

Other oxide materials of possible interest for their combination into ICs are aerogels. Aerogels are reticulated polymers obtained by a process of hydrolysis and

---

[8] The ultimate reason for the differences between man-made and biosystems is perhaps related to the objects used for managing the information, actuate the action, or elaborating the signal coming from the external world: Man has given his preference to electrons, whereas Nature has preferred ions and molecules. Understanding the reason for Nature's preference is not trivial. We speculate that it is ultimately related to the need for generating ordered structures able to perform even sophisticated operations in a chaotic medium. The kinetics of chemical networks may indeed exhibit computational functions; in particular, spatially homogeneous networks (inspired to reactions occurring in living cells) show the behaviors of logic gates and sequential and parallel networks [275]. Open systems subjected to a sufficiently high power income may self-organize, both spatially and temporally, concentrating a large number of degrees of freedom offered by the thermal reservoir in a relatively narrow region of the phase space. This is however possible only for objects whose degrees of freedom may be excited thermally – for ions and molecules but not for electrons.

polycondensation of suitable monomers in a nonreacting solution and the elimination of the solvent after reaction completion. Considering, for instance, a solution of $Si(OR)_4$ in ROH (with $R = CH_3, C_2H_5$, etc.), the addition of the first molecules of water leads to the following reactions:

$$2(RO)_3SiOR + 2H_2O \longrightarrow 2(RO)_3SiOH + 2ROH \qquad \text{(hydrolysis)}$$
$$\underline{\ 2(RO)_3SiOH \longrightarrow (RO)_3SiOS(OR)_3 + H_2O \quad \text{(condensation)}}$$
$$(RO)_3SiOR + H_2O \longrightarrow (RO)_3SiOS(OR)_3 + 2ROH$$

The addition of further water results in the hydrolysis of other monomers and of other terminations of the condensed oligomer so that the completion of the process will produce a reticulated polymer resulting from the diffusion-limited aggregation of the various reactants, embedded in the original solvent and embedding part of it. Removing the solvent by evaporation produces the collapse of the delicate silica gel structure because of the raise of the same capillary effects as considered in Sect. 8.3. It is however possible to cancel such effects transforming the whole liquid into vapor via supercritical extraction. What remains after the elimination of the solvent as a vapor is a polymer dispersed in the entire region originally occupied by the solution – an *aerogel* [276–278]. It was just with this method that the aerogels were first produced [279]. Currently, several methods and solutions are used: the supercritical extraction of alcohol is unsafe (because of explosion risk); substituting water for alcohol removes the risk of explosion but requires harder (pressure and temperature) processing conditions; the substitution of CO_2 for water or alcohol softens the extraction conditions but is poorly consistent with large aerogels [280]. Removing of the solvent without the use of supercritical extraction from thin films coating solid surfaces is possible [281], but requires an accurate control of the hydrophobicity–hydrophilicity properties of the polymer side chain on the same basis of the considerations of Sect. 8.3.

The aerogel structure is very open and may be viewed as a fractal[9] [282]. If the apparent density of the aerogel is sufficiently low, the structure is so open as not to manifest any appreciable surface tension; this is confirmed by the fact that the densification of the aerogel to the bulk density occurs preserving the original shape (the same as that of the vessel where it was synthesized). The preparation of aerogel samples with even complex shapes is possible with the use of a suitable mould; the densification of the sample to the bulk density can be stopped to preserve the original shape. Figure 8.17 shows that shape may be preserved for complex geometries too.

Having in mind the last property, the aerogel technology can be applied for the preparation of net- or near-net-shaped materials for optical applications [283]; the rigid silica skeleton of the aerogel can instead be used to impart stability to

---

[9] The reader is assumed to have an at least qualitative idea of what is a fractal; for more information see Chap. 10.

Fig. 8.17 Silica monoliths prepared via sol-gel after the synthesis (*middle*), supercritical extraction of the solvent (*left*), and densification (*right*)

delicate organic substances (like enzymes or antibodies) without destroying them [284–286]; the rigidity of the skeleton and the openness of the structure may be used for the preparation of highly energetic species otherwise unattainable with the usual synthetic routes [287, 288].

Concluding Remarks

The prediction that

> the development of hybrid architectures, molecules working in concert with silicon, is among the most likely paths that this technology will take in the coming years as we develop ultradense and ultrafast computational systems

is certainly not new [10, p. 251]. Modifying the silicon technology in such a way that silicon-based circuits can host billions of batch-fabricated molecular devices forming the hybrid circuitry is however a major technological problem.

This problem has been attacked separating it into four parts:

- The setup of an economically sustainable technology for the preparation of TSI crossbars
- The demultiplexing of the addressing lines to allow their linkage to the CMOS circuit
- The design and synthesis of the appropriate functional molecules
- The grafting via batch processing of the functional molecules to nanoscale crosspoints

Solutions have been found for each part. The combination of these solutions for the realization of demonstrators of hybrid TSI circuits in silicon seems now possible through a *conservative extension of the current CMOS technology*.

In principle this technology, still in nuce, seems also able to give access to the exploitation of molecular motors and to the functional study of biological systems at the cellular and subcellular level. The second part of this book is targeted at these topics.

Part II
Advanced Topics: Self-Similar Structures, Molecular Motors, and Nanobiosystems

Chapter 9
Examples

Although single solutions to the problems formulated at the end of Chap. 4 are known, combinations suitable for the preparation of a true hybrid molecular circuit are still missing. The following part is devoted to speculate on a few possible perspectives, listed and discussed in order of increasing content of molecular properties at circuital level.

9.1 Hybrid Molecule–MOS-FET Combination

Features with characteristic size on the nanometer length scale are already produced in CMOS processing [106], even unintentionally. Figure 9.1 shows an example where the top poly-Si film is separated from the underlying single-crystalline silicon by an SiO_2 layer with thickness comparable with molecular length.

Assume that the poly-Si film works as floating gate of a flash memory [289]. In these devices the insulating layer separating the gate from the underlying channel behaves as a strongly nonlinear resistance: an almost ideal insulator in sensing mode (very low I^{OFF}) and as good conductor during writing or erasing (very high I^{ON}). The characteristic times for writing or erasing $t_{W/E}$ and the retention time t_{ret} decrease with I^{ON} and I^{OFF}, respectively, so that $I^{ON}/I^{OFF} = t_{ret}/t_{W/E}$. For $t_{W/E}$ on the millisecond time scale and t_{ret} of a few years, the oxide region separating floating gate from drain or source behaves as a nonlinear resistor with $I^{ON}/I^{OFF} \approx 10^{11}$. This value is exceedingly large and shows how the current IC technology has succeeded in engineering the structure to impart properties not available in any material.[1]

The insertion of a molecule in the recessed region of Fig. 9.1 would greatly simplify the operation mode of NVMs, but the design and synthesis of molecules with $I^{ON}/I^{OFF} \approx 10^{11}$ seems a hopeless dream. The use of reprogrammable molecules

---

[1] To give an idea of how large the ratio I^{ON}/I^{OFF} is, consider that degenerate n^+ silicon has resistivity ϱ_{n+} of the order of 10^{-3} Ω cm, whereas the resistivity of undoped silicon ϱ_v is typically 10^6 Ω cm, so that $\varrho_v/\varrho_{n+} \approx 10^9$. Degenerate n^+ and p^+ silicon materials have similar resistivities.

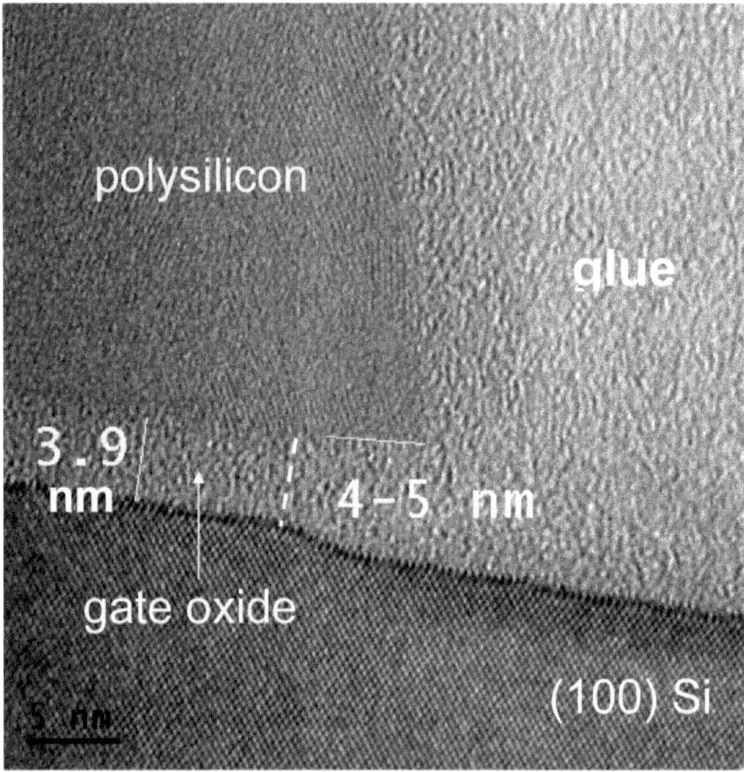

Fig. 9.1 Cross section image at the transmission electron microscope of a nanometer-sized region unavoidably produced in CMOS processing. The recessed region (having the gate oxide, poly-Si, and single-crystalline silicon as borders, and occupied by a glue to permit a better contrast in transmission electron microscopy) may be used to host molecules able to perform electrical functions. Lines have been drawn to guide eyes

as memory elements of static memories would imply a lifetime on a time scale of 10^4 s and hence $I^{ON}/I^{OFF} \approx 10^7$, still too high for currently achievable molecules. The exploitation of reprogrammable molecules in memories may thus be done only in applications, such as dynamic RAMs (DRAMs), where the information is periodically refreshed.

Realistic values of I^{ON}/I^{OFF} are of the order of 10^5 (pentacene-based plastic transistors with I^{ON}/I^{OFF} higher than 10^5 at operating voltage as low as 5 V have been reported [290]). This value is incompatible with the applications considered above, but might allow the practical use of the molecules as a protection against electrostatic discharge in conventional (rather than floating-gate) MOS-FETs to link the gate to source or drain.

9.2 Crossbar Functionalization

Following [291] closely, in the forthcoming part the case of a molecule with rose-bengal core will be considered. Fluorescein derivatives (see **1** in Fig. 9.2 for its sodium salt) are known to display, in the bulk, electrical bistability; in particular, rose bengal (**2** in Fig. 9.2 [292, 293]) has $I^{ON}/I^{OFF} \approx 10^5$ and tolerates 10^6 writing–reading cycles the operating voltages being around ± 3 V [294–296].

The electrical bistability is attributed to the formation of an anion radical stabilized by the presence of electron withdrawing halogen groups (from **3** to **4** in Fig. 9.3) [294] . The reduction produces the aromatization of the quinonic ring and hence an increase of conductivity [294–299].

Actually, there is not a general consensus about the bistability of rose bengal: recent studies [300] suggest indeed that the bistability observed in devices formed by a (multimolecular) layer of rose bengal embedded between metal electrodes is better understood in terms of interfacial redox phenomena. Without subscribing a priori to any of these two interpretations and mainly for concreteness reasons, in the following the case of rose bengal will be assumed, although the considerations of Sect. 8.3 extend to any polar molecules with nonpolar arms.

In this part the attention is focused on molecules **5** and **6**, characterized by a core derived from fluorescein and by two arms terminated with alkyne tails. In its reduced

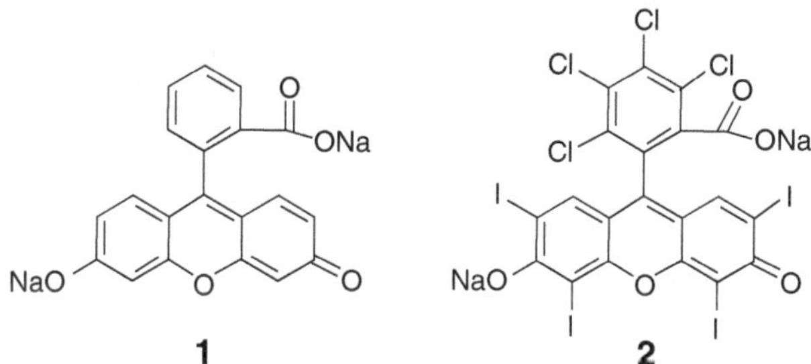

Fig. 9.2 Sodium salts of fluorescein, **1** and of rose bengal, **2**

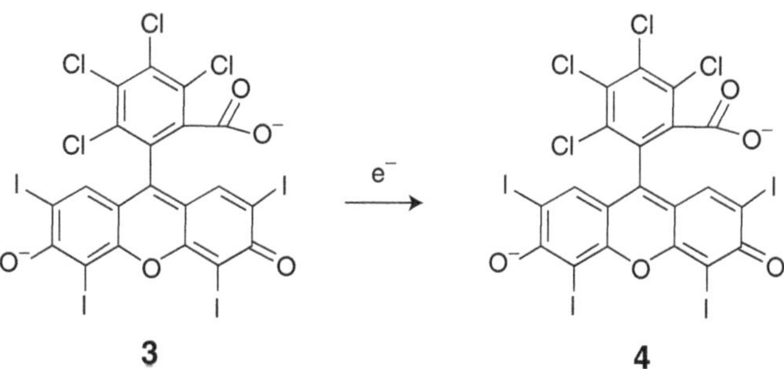

Fig. 9.3 Reduction of rose-bengal dianion **3** with formation of a π-conjugated core **4**

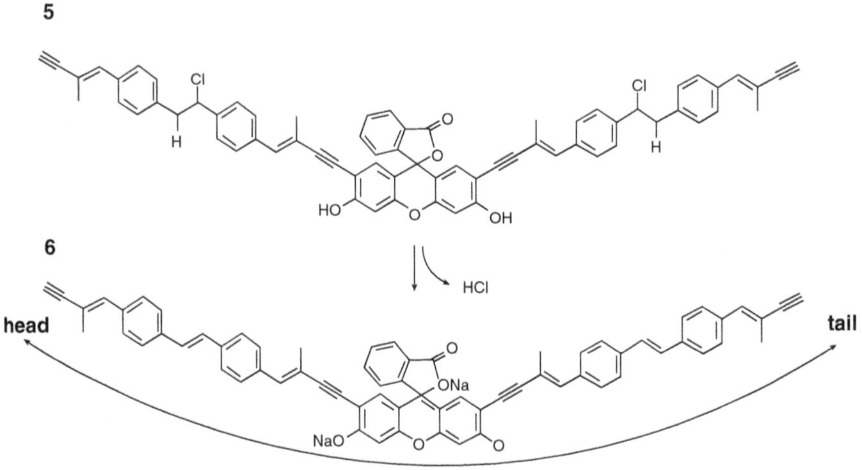

Fig. 9.4 Formation of the functional molecule with π-conjugated arms via elimination of sacrificial groups, **5 → 6**

form, the complete π conjugation of **5** is destroyed by the σ bonds linking carbon atoms with X and H terminations (X = Cl, Br, I); the elimination of these groups (**5 ⟶ 6**), catalyzed by the presence of Brønsted bases, restores the complete π conjugation (Fig. 9.4).

Synthetic routes for the preparation of molecules like **5** can be hypothesized: A possible synthesis strategy exploiting commercially available products involves three steps:

- Synthesis of tails
- Chemical linkage of the tails to the fluorescein core
- Conversion of fluorescein lactonic form to electrobistable acid one

In particular, the first step starts from the hydrohalogenation of *trans*-stilbene ([(*E*)-2-phenylethenyl]benzene) followed by the bromination of the product. The brominated species is then coupled with 2-methyl-1-buten-3-yne by the Heck reaction [301]. In the second step the fluorescein core is linked to the tails prepared as above by the Sonogashira reaction [302]. At last the fluorescein derivative is converted to the active acid form by suitable change of the solution pH. Figure 9.5 sketches the three steps.

The equilibrium configurations of molecules **5** and **6** are sketched in Fig. 9.6. That their geometries really satisfy the conditions that allow the insertion of **5** in the recessed region and the grafting of **6** to the silicon surfaces, is confirmed by distributions of their tail-to-head distance at a temperature, 500 K, consistent with reaction **5 ⟶ 6**. Figure 9.7 shows that the head-to-tail distance in **5** has a distribution with

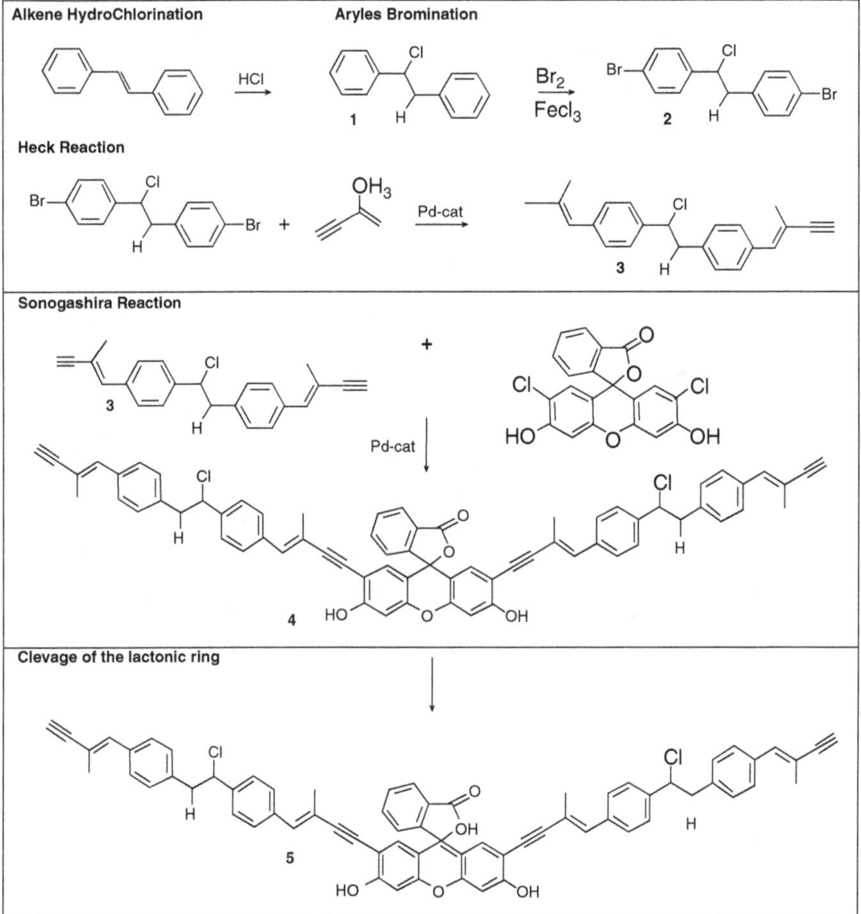

Fig. 9.5 Sketch of the synthetic route for the preparation of functional molecule **6**

mean value of 2.1 nm and standard deviation of 0.5 nm, whereas in **6** the mean head-to-tail distance is 3.6 nm with a standard deviation of 0.3 nm. This is consistent with the idea that the increased conjugation puts the molecule in an almost planar conformation.

It is noted that the distance distributions shown in Fig. 9.7 are fully consistent with the possibility of preparing a sacrificial oxide layer adapted to accept molecular grafting.

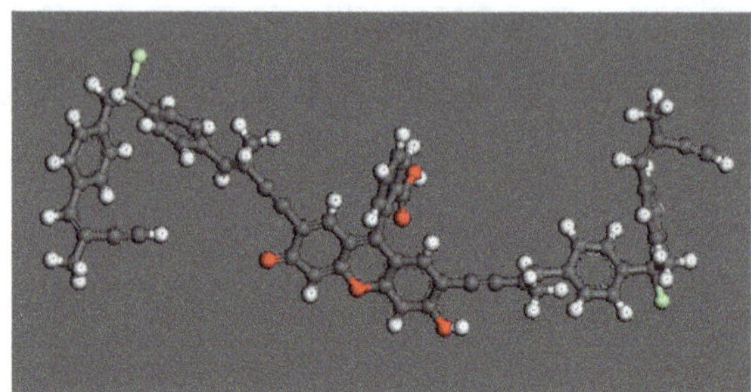

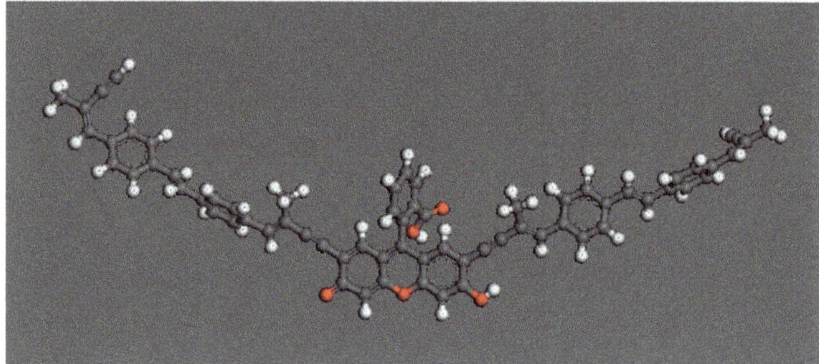

Fig. 9.6 Stick-and-ball picture of the equilibrium configurations of the molecules with broken (*up*) and full (*down*) π conjugation

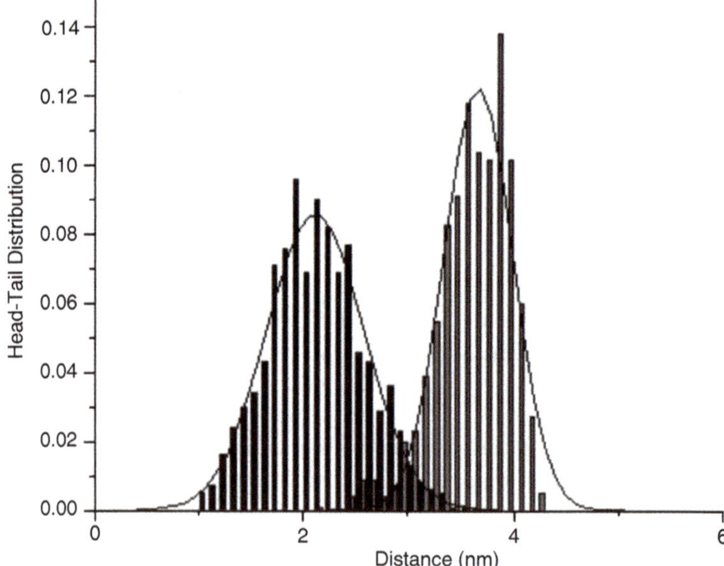

Fig. 9.7 Histogram of the head-to-tail distance at 500 K in precursor with σ bridges, **5** with X = Cl (*dark bars*), and in the molecule with full π conjugation, **6** (*clear bars*). The distances were calculated simulating the molecular dynamics with Material Studio by Accelrys, using Discovery module

Chapter 10
Self-Similar Nanostructures

The geometric description of the world is at the root of Western thought: to the first (perhaps mythic) philosopher, Thales, we owe a theorem of plane geometry. The role of geometry in the construction of physical theories is equally important: on the cosmological scale, gravitation is explained in terms of space curvature; on the submicroscopic scale, quantum mechanics may be explained in terms of fluctuation of the metric tensor [303].

On the laboratory length scale, the most important geometric property of any thermodynamic system is most likely its shape. The general mathematical description of surfaces is a notoriously complicated affair – especially if one tries to describe real surfaces.

10.1 Fractals

Fractals are self-similar objects. Pathological objects with self-similar characteristics were already known to mathematicians at the end of the nineteenth century (Pean and Koch curves, Cantor middle-excluded set, etc.); however, fractals took on a central role in the description of the physical system much later, with Mandelbrot's observation of the scale invariance of several physicochemical systems [304].

10.1.1 Queer Systems

The area A of bodies with a *regular shape* (such as the cube, sphere, polyhedra, etc.) increases with volume as $V^{2/3}$. For bodies with a regular shape, the effect of the surface on their intensive thermodynamic properties (such as specific heat, magnetic susceptibility, etc.) disappears in the thermodynamic limit (infinite volume at constant density).

A body whose area increases with V faster than $V^{2/3}$ is said to have a *queer shape*. The area of a queer system is well defined; its effect on intensive thermodynamic properties, however, can persist even for $V \to \infty$. The effect surely persists in the thermodynamic limit for bodies with $A \propto V$.

Though it may appear that the concept of a queer system is an extravagant concept, Nature abounds in bodies with queer shape (the first systematic analysis for queer systems was recorded in [305]). Among the most interesting queer systems are zeolites (in which queerness is due to a lattice of void cages connected by tubes, regularly arranged in the system), biological systems (because reproduction is a way which allows the overall area to increase in proportion to the volume), and films obtained by low-temperature PVD (in which the condensed film grows with an exposed number of sites increasing in proportion to the average thickness [306]).

Queer systems, however, are not sufficiently extravagant to exhaust the possibilities of Nature – fractals are even more extravagant.

10.1.2 Fractals in Mathematics

That smoothness is not a mandatory feature of the way Nature expresses itself, but rather, fractality is of ubiquitous occurrence in a large class of phenomena even in the Mineral Kingdom, has been clear ever since the genesis of the theory, with Mandelbrot's question about the length of Great Britain's coast [304].

From a mathematical point of view, a fractal set exhibits the property that the "whole" can be represented as the collection of several parts, each one obtainable from the "whole" by a contracting similitude [304]. A typical fractal object is self-similar, i.e., a magnified portion of it appears identical to the entire object observed under lower resolution: from this point of view it is said to be invariant under scale transformation. If the contraction similitude occurs in less than three dimensions, the object is said to be self-affine.

A fractal object can usually be defined through a successive iteration process in which an initiator is contracted with a similarity ratio $1/\chi$ and put $v(\chi)$ times in a given arrangement called generator, the same operation then being repeated again and again without end. If the area of a fractal set varies with the probe yardstick ρ as in

$$A \propto \rho^{2-D}, \tag{10.1}$$

(which diverges for $D > 2$), the fractal dimension D of the set is given by[1]

$$D = -\lim_{\chi \to \infty} |\ln v(\chi)/\ln(1/\chi)|.$$

[1] For any set X in $\mathbb{R}^n$ its dimension, $\dim(X)$, is a positive number with domain $[0, n]$ satisfying the following properties [307]:

- (*Limiting behaviors*) for the singlet set $\{P\}$, $\dim(\{P\}) = 0$; for the interval I, $\dim(I) = 1$; for the hypercube I^m ($m \leq n$), $\dim(I^m) = m$
- (*Monotonicity*)

$$X \subseteq Y \implies \dim(X) \leq \dim(Y)$$

Fig. 10.1 Looking at cauliflower details at different magnifications

10.2 Fractals in Nature

To be convinced that self-similarity is not so extravagant, it is sufficient to observe the cauliflower at different magnifications (see Fig. 10.1).

10.2.1 Fractal Biological Systems

Life can be preserved against the Second Law of thermodynamics only in the presence of a production of negative entropy in proportion to the total mass of the organism. This is achieved via the establishment of a diffusion field sustained by the metabolism inside the organism [308].

For any regular shape, a body of volume V admits an area A satisfying the following condition:

$$V = g_3 A^{3/2}, \qquad (10.2)$$

- (*Countable stability*)

$$\dim\left(\cup_{j=1}^{\infty} X_j\right) = \sup_{j \geq 1} \dim(X_j)$$

- (*Invariance*) for any arbitrary map Ψ belonging to some subfamily of the set of homeomorphisms of $\mathbb{R}^n$ to $\mathbb{R}^n$

$$\dim(\Psi(X)) = \dim(X)$$

A dimension can be assigned to subsets of $\mathbb{R}^n$ in several in general, not coinciding, ways. The coverage with balls is one of them.

with g_3 being a coefficient related to the shape: $g_3 = (1/6)^{3/2}$ for cubes, $g_3 = 1/6\sqrt{\pi}$ for spheres, etc. The well known variational property of the sphere, of embedding the maximum volume for assigned area, guarantees for all bodies: $g_3 \leq 1/6\sqrt{\pi}$. In bodies of regular shape, the ratio A/V diverges for small V and vanishes for large V.

As the consumption of energy and negentropy increases in proportion to V whereas the exchange of matter increases A, small unicellular organisms (like bacteria) have no metabolic problem due to their shape and can grow satisfying (10.2), preserving highly symmetric shapes; actually, the smallest living bacterium (the pleuropneumonia like organism with diameter 0.1–$0.2\,\mu$m) is spherical. On the contrary, larger organisms (like the amoeba) may survive only by adapting their shapes to have an energy uptake coinciding with what is required to preserve vital functions [305, 309]:

$$V = g_2 A, \qquad\qquad (10.3)$$

with g_2 being a coefficient characteristic of two-dimensional growth. Moreover, the need to match the variable environmental conditions is made possible only because of the existence of an inner organelle, the endoplasmatic reticulum [310], which can fuse to the external membrane, thus allowing an even larger change of area at constant volume [305, 309]. This mechanism, however, can sustain the metabolism of unicellular organisms of the size of, at most, $10\,\mu$m (amoeba).

Larger organisms require the organization of the constituting cells in tissues specialized for single functions. This specialization, however, requires the formation of a vascular network (to transport anabolites to, and catabolites from, each constituting cell) whose need to be dense inside the organism impart to the network a space-filling nature. Although a space-filling system is not necessarily a fractal,[2] that is, indeed, the case for a vascular system [312]. Nature prefers to manifest itself with fractal shapes in other situations too, like in the mammalian lung [313] or in the dendritic links of the neuron (see Fig. 10.2).

10.2.2 Fractal Surfaces

When the geometrical irregularities in the surface have spatial extensions which are comparable with the size of the adsorbed molecules, new phenomena are expected to occur. In particular, several surfaces observed at a length scale 3–15 Å give evidence for fractality.

For fractal surfaces, the concept of surface area loses its original absolute meaning and becomes relative to the probe through which the area is determined. The surface area of a solid can be determined by choosing a probe gas, with effective cross section b_2° (known, for instance, from the molar volume of liquid

---

[2] An example is given by artificial dendrimers. A dendrimer is a branched oligomer, whose growth is eventually limited by space filling [311]. The dendrimer is not scale invariant and thus is not a fractal.

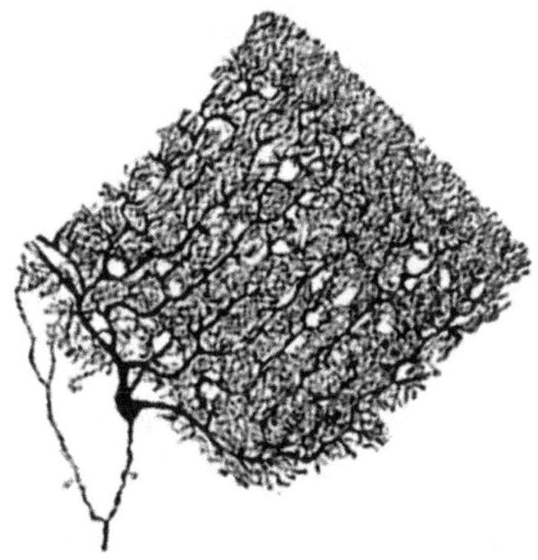

Fig. 10.2 Micrograph of a human Purkinje cell stained to show its dendritic structure

adsorbate), determining with adsorption techniques, the number $\mathcal{N}_m^\circ$ of molecules forming a monolayer [e.g., as results from the adsorption isotherm plotted according to the Brunauer–Emmett–Teller (BET) equation [314]], and taking the product $A^\circ = \mathcal{N}_m^\circ b_2^\circ$. The area determined by the BET plot of the N_2 adsorption isotherm at 77 K in the BET confidence range has become a widely accepted standard.

This procedure would not produce ambiguities if such a product for a different gas, $A = \mathcal{N}_m b_2$, were independent of the gas itself. Actually, the relationship

$$\mathcal{N}_m = \mathcal{N}_m^\circ (b_2^\circ / b_2) \tag{10.4}$$

is found to be satisfied only by certain adsorbents [315]. Adsorbents that are not described by (10.4) are usually found to obey the equation

$$\mathcal{N}_m = \mathcal{N}_m^\circ (b_2^\circ / b_2)^{D/2}, \tag{10.5}$$

with $2 < D < 3$ [315]. Defining the characteristic length ϱ of the probe as $\rho = \sqrt{b_2}$, surfaces whose area varies with ρ as in (10.1) satisfy (10.5), so that they are fractal, with D being their *fractal dimension* [316].

The identification of an irregular surface with a fractal set poses strict constraints on the surface characteristics: it implies the recurrence of the same irregularity details when the surface is magnified successively. Although this seems to be a very limitative condition, the fractal behavior of adsorbing surfaces is remarkably abundant [315, 317–319].

The fractal nature of an adsorbing surface has important consequences on its adsorption behavior [320–322]. Indeed, any adsorbed molecule has a number of available adsorption sites between those which are characteristic of two and three dimensions, and this fact affects their statistico-mechanical behavior. Such an effect is even more pronounced when lateral interactions between adsorbed molecules are taken into account.

The influence of the adsorbent fractal geometry on the adsorption isotherm has been taken into account along two general approaches: statistical mechanics (where the fractal nature of the surface is allowed for considering how geometry affects the partition function [320]) and kinetics (where the effect of geometry on the adsorption and desorption rates is considered [322]).

An absolute understanding of adsorption on fractal surfaces has not been achieved yet, and such studies constitute an active area of research. In particular, little is known with regard to the adsorption on simultaneous geometrically (fractal) and energetically heterogeneous surfaces. Nevertheless, even maintaining the usual scheme employed in the study of energetically heterogeneous surfaces presented above, it is easy to forecast substantial alterations of the local isotherm, especially when lateral interactions are considered.

Fractal surfaces are obviously obtained under strong nonequilibrium conditions. This property leads to identification of electrochemistry (anodic oxidation leading to dendritic corrosion) or sol-gel technology (SiO_2 gels, especially via acid catalysis, in which filamentary, weakly branched, structures are produced [280]) as candidates for the preparation of fractals.

As fractal surfaces are nonequilibrium configurations, the difference $D - 2$ can be seen as a generalized driving force toward equilibrium. Because adsorption is a way to restore equilibrium at unsaturated surface bonds, adsorption is presumably a way to reduce dimensionality. Surface defractalization by adsorption has actually been observed [321].

The fractality of Nature is obviously approximate, the lowest length scale being ultimately limited by the atomic nature of matter. That surfaces may continue to have a scale invariance down to the atomic size became clear, however, only after Avnir, Farin, and Pfeifer's study of the adsorption behavior of porous adsorbents [315–317, 323].

10.3 Fractals in Technology

It is relatively easy to prepare a fractal pattern designing it at the human scale and reproducing it at a lower scale via photolithography. The lowest length scale consistent with this process is the lithographic limit. Extending fractality to a lower length scale (thus producing nanofractals) is not simple.

Imagine for a moment that, in spite of the atomistic nature of matter and of the inherent technological difficulties, the multiplicative route can be repeated indefinitely. Remembering that the $(n + 1)$th step generates a set S_{n+1} that is nothing but the one at the nth, S_n, at a lower scale, the sequence $\{S_0, \ldots, S_n, \ldots\}$ defines a fractal. The fractal is self-similar only if the height of each spacer varies with n as 2^{-n}; otherwise, the fractal is self-affine [324]. As discussed for the additive route, the height t_n decreases "spontaneously" in an almost linear fashion with n, $t_n = t_0 - n\tau$ (5.9), so that the fractal is self-affine. If τ is negligible, a self-similar fractal can be obtained at the end of process planarizing the whole structure with a

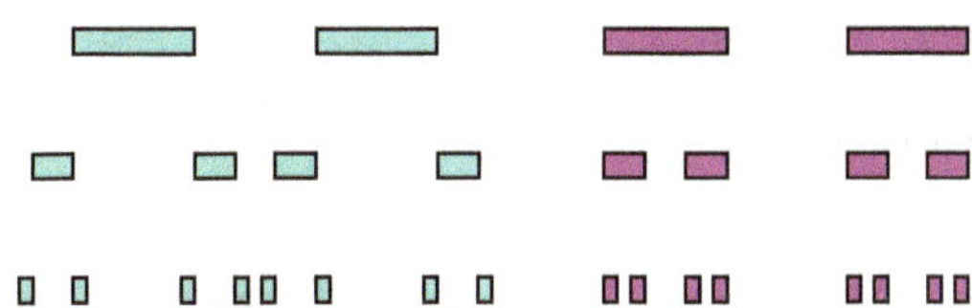

Fig. 10.3 Generation of the multispacer set (*left*) and of the Cantor middle-excluded set (*right*)

resist and sputter etching the composite film in a nonselective way until the thickness is reduced to $t_0/2^n$. Even ignoring the technological factors, the atomistic structure of matter limits the above considerations to an interval of 1–2 orders of magnitude, ranging from a few atomic layers to the lower limits of standard photolithography.

Having clarified in which limits the set S_n may be considered as a fractal, it is interesting to compare it with other fractal sets. The prototype of such sets, and definitely the most interesting from the speculative point of view, is surely the Cantor middle-excluded set. Figure 10.3 compares sequences of three discrete processes eventually leading to the multispacer fractal set S and to the Cantor set C. The comparison shows interesting analogies: Take $P = 2W$; if $w_n = (1/3)w_{n-1}$, the measure of each multifractal-step set coincides with that of the Cantor-step set. This implies that the multispacer fractal set has null measure. Similarly, it can be argued that the multispacer set, considered as a subset of the unit interval, has the same fractal dimension as the Cantor middle-excluded set – $\ln(2)/\ln(3)$ [324].[3]

At each step, the multispacer fractal set is characterized by a more uniform distribution of single intervals than the Cantor set; this makes the former more interesting for potential applications than the latter. In spite of this, trying to reproduce the Cantor set on the nanometer length scale seems to be of some interest. It is possible with existing technologies. Following closely [325], a possible process for this involves the following steps:

1. The lithographic definition of the seed (formed, for instance, by poly-Si) generating the Cantor set
2. Its planarization (for instance, via the deposition of a low viscosity glass and its reflow upon heating)
3. The etching of this film to a thickness controlled by the exposure of the original seed
4. The selective etching of the original film
5. The conformal deposition of a film of the same material as the original seed (poly-Si, in the considered example) and of thickness equal to 1/3 of its width
6. Its directional etching
7. the selective etching of the space seed (glass, in the considered example)

Figure 10.4 sketches the overall process.

The preparation of fractal structures may appear at first sight to be nothing but a mere exercise of technology stressing. However, an important point is stressed: if the

---

[3] Note that the multispacer set obtained with $\varrho = \frac{1}{2}$, see equation (5.7), is indeed a self-similar set, but it is not a fractal, and its dimension is exactly 1.

Fig. 10.4 A process for the generation of Cantor's middle-excluded set

Fig. 10.5 Plan-view comparison of the crossbars obtained crossing two multispacer fractal sets (*top*) and Cantor middle-excluded sets (*bottom*)

S^nPT is used for the preparation of crossbar structures for molecular electronics, the functionalization of the crosspoints with organic molecules can only be done after the preparation of the hosting structure. As discussed in Sect. 8.3, this requires an accurate control of the rheological and diffusion properties in a medium embedded in a domain of complex geometry. Understanding how such properties change when the size is scaled and clarifying to which extent the domain can indeed be viewed as a fractal (allowing the analysis on fractals [326, 327] to be used for their description) may be a key point for the actual exploitation of already producible nanometer-sized wire arrays in molecular electronics.

Figure 10.5 shows a plan-view comparison between a 16×16 crossbar obtained via $S^3PT_\times$ starting from lithographically defined seeds separated by a distance satisfying (5.8), and a 16×16 crossbar obtained via $S^3PT_\times$ starting from lithographically defined seeds and arranging the process to generate the Cantor middle-excluded set. To draw the figure, the lithographic widths have been chosen to allow the production of sublithographic wires with width (12.5 nm) that has been proved to be producible [124].

Chapter 11
Molecular Motors

Macroscopic motors are usually characterized by their ability to transform the input energy into mechanical work via the generation of a force, in turn responsible for an acceleration, eventually resulting in a velocity, limited by the unavoidable friction (Fig. 11.1). The same principle extends to microelectromechanical systems (MEMSs) which are nothing but macroscopic motors scaled to the micrometer length scale and produced with techniques borrowed from the CMOS technology. Macroscopic machines as well as MEMSs are dominated by inertia: the time required by friction to put the particle at rest increases with the mass of the moving part and is typically on the laboratory time scale.

For bodies of molecular size this time becomes vanishingly small, on the picosecond time scale, and in this case the generation of mechanical energy with the scheme of Fig. 11.1 is not convenient. Rather, the most convenient way is the direct transformation of the input energy into mechanical work via rectification of the Brownian motion responsible for friction (Fig. 11.2). Systems of this kind are often observed in biosystems.

So far the attention has been concentrated on the conduction properties of molecules whose conductance can be varied via external or internal redox processes induced by electrical activation. Although the change of electronic states produced by these processes certainly involves conformational changes, the role of conformation seems limited to provide metastability to the excited electronic state.

In recent years a lot of work has been spent on the design and synthesis of molecules displaying large positional displacement of submolecular units in response to energetic stimuli. These molecules are usually referred to as molecular "motors" or "mechanical machines" and are usually named as the corresponding macroscopic machines exhibiting the same functions.[1] A lot of molecules able to perform even complex actions, are actually known: for instance, "molecular rotors" [328, 329], a "molecular elevator" [330], and a "nanocar molecule" [331] have

---

[1] The following definitions have been accepted: A *device* is any system designed to perform an assigned function. An *engine* is any device for the transformation of a form of energy (e.g., mechanical, thermal, chemical, optical, etc.) into another form (e.g., mechanical or electrical), usually more concentrated or with lower entropy content. A *motor* is any device for the transformation of a form of energy (e.g., mechanical, thermal, chemical, optical, etc.) into mechanical work.

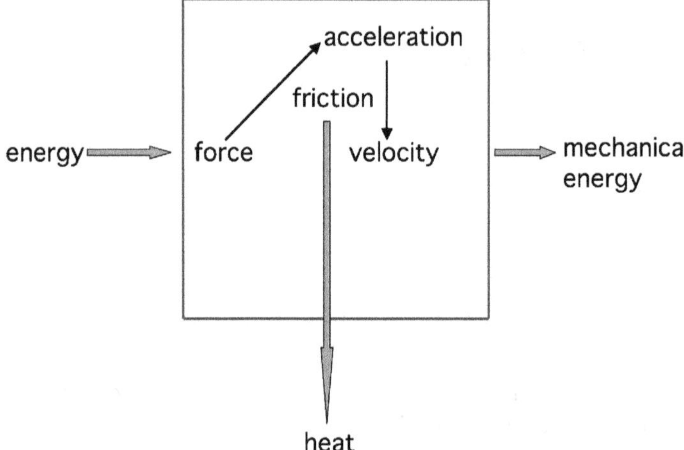

Fig. 11.1 Block diagram of a macroscopic motor

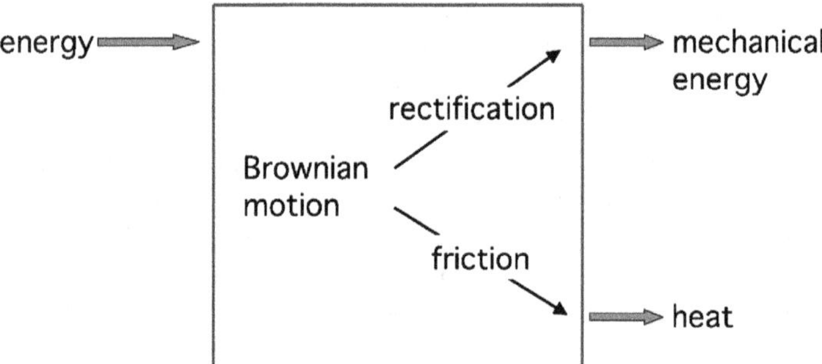

Fig. 11.2 Block diagram of a Brownian motor

been described. Extended reviews are given, for instance, in [332–336]; a friendly trip in the magic world of molecular machines is proposed in [337].

In most cases, these molecules or supramolecular complexes are operated in solutions or, more rarely, at the surface of solids. In most cases, although the actuation of molecular machine is made by externally controlled stimuli (e.g., lighting), the work they make cannot be exploited externally.[2] Really few exceptions are known [338–

---

[2] Though such molecular motors or engines are often described in thermodynamic terms, in the author's opinion this is allowed only if the mechanical or electrical work they produce is externally accessible. The possibility of a thermodynamic description (and hence the possibility of considering a supramolecular complex as a system characterized by internal conformational degrees of freedom, although embedded in a medium) seems thus related to their grafting to externally accessible contacts. Of course, the concept of "accessibility" is related to the nature of the observer: thus the work done by a myosin complex immersed in a solution containing adenosine triphosphate (ATP) may be not accessible by a human observer but exploited by a bacterium of which it forms its flagella.

340]; in these cases the molecules may really be viewed as molecular motors or engines.

When the operation is carried out in solution without selecting any special molecule and trying to extract the work done, the most convenient tool for the actuation is lighting [337]. The situation changes when one tries to actuate selected molecules or to exploit the work externally. In this case the most convenient form of energy is the electrical one. When the work is made under an electrical stimulus applied between two or more externally accessible nanoscopic contacts (or, vice versa, electrical energy is generated from mechanical work) the molecule or a supramolecular complex behaves as a genuine *molecular electromechanical system* (mEMS).

This chapter is targeted at the description of highly idealized examples of molecules inserted between contacts and working as mEMS.

11.1 Molecular Building Blocks

To allow translational or unrestricted rotational motion, the molecular systems are necessarily formed by supramolecular assemblies [341] and are generally understood in the frame of topochemistry [342–345]. Among the numerous supramolecular systems, rotaxanes play a central role.

A rotaxane (A, R) is a supramolecular system formed by a cyclic molecule R ("ring") threaded on a linear molecule A ("axle") with bulky caps on the heads to prevent the ring from falling off (see Fig. 11.3).

Chemically speaking the ring is usually a rigid, nearly flat, arene (a famous example being the "paraquat" – two doubly charged bipyridinium units linked by two benzene rings, with the net positive charge balanced by four PF_6^- anions), whereas the axle is formed by n sites (for instance, benzene units) typically connected by chains of ethoxy groups.

Rotaxanes can be synthesized with different strategies, named according to the last stage of the synthesis: in *capping* the last stage is the reaction of axle terminations with bulky groups and is made possible by the strong noncovalent interactions that stabilize the axle when threaded through the ring; in *clipping* the last stage is the closure of the portion of ring (partial macrocycle) noncovalently stabilized around the axle; while in *slipping* the size of end groups is small enough to allow the dumbbell molecule to thread through the ring at high temperature but so large as to render impossible the escape when the temperature is lowered.

Rotaxanes are almost ideal building blocks for mEMSs because they can be used to support controlled translational or rotational motion (see Fig. 11.4).

Up to now, rotaxanes have mainly (or perhaps uniquely) been hypothesized or used because of the possibility of functionalizing the axle via the insertion of two or more positions where the ring is preferentially hosted [96] (actually, are just these sites the ones that make possible capping and clipping). The reader interested in this topic may refer to a vast literature [332–337].

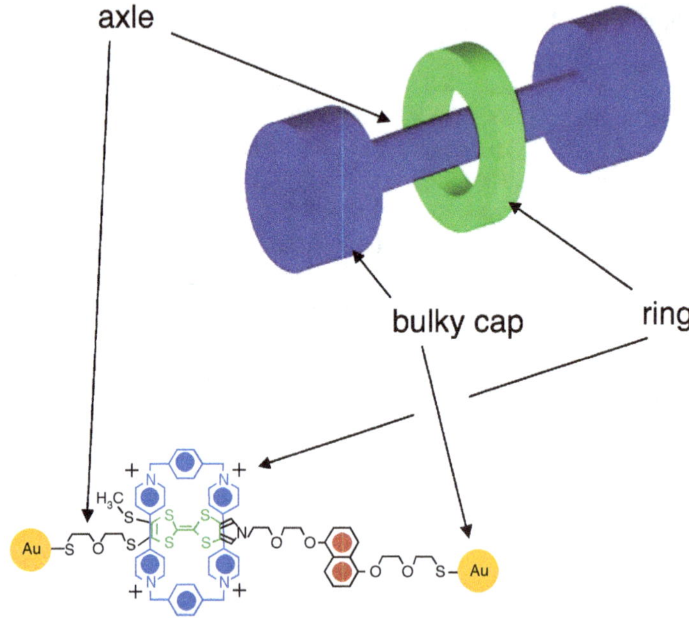

Fig. 11.3 Schematic view of a rotaxane and its actual realization (capping is done with gold nanoclusters)

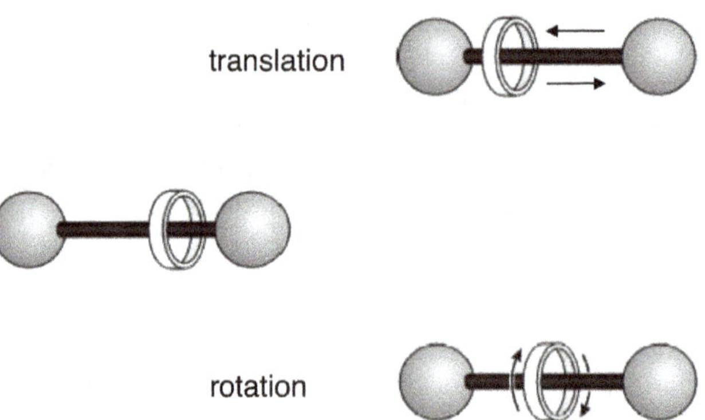

Fig. 11.4 Exploiting rotaxanes for their ability to impart controlled translational and rotational motions

However, there is another possibility that could be exploited – the functionalization of the ring. This possibility is of special interest because it allows the role of deterministic external forces and erratic internal forces to be clarified.

For that, the attention will be concentrated on the following ideal configuration:

- The axle is an unstructured, rigid, nearly linear, insulating wire connecting two electrodes C_1 and C_f.

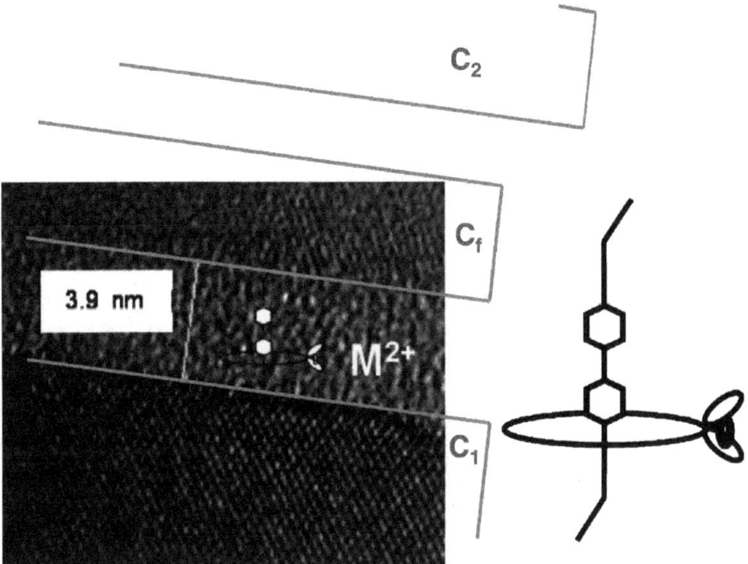

Fig. 11.5 The considered molecular motor: a rotaxane, functionalized via the insertion of a transition metal in the ring admitting two oxidation states, is grafted between silicon electrodes C_1 and C_f. One of them (say C_f) is electrically floating and its potential is controlled by an outer electrode C_2

- C_f is electrically floating and geometrically arranged between C_1 and another electrode C_2.
- A potential difference may be applied between C_1 and C_2 so that C_f attains an intermediate potential.
- The ring is suitably functionalized to admit several charge states.

The considered arrangement is shown in the left-hand side of Fig. 11.5, while the hypothesized molecule is sketched in the right-hand side. Two different "ring engineering" arrangements will be considered in Sects. 11.2 and 11.3.

11.2 Controlling Movement with Electric Field

Assume that the ring is negatively charged and that the transfer of the extra electron of R^- to electrodes C_1 and C_f is thermodynamically forbidden. This condition holds true when the energy level of R^- [$-A_{el}(R^0)$, where $A_{el}(R^0)$ is the electron affinity of R] is appreciably lower than the Fermi energies $E_F(C_1)$ and $E_F(C_f)$ of C_1 and C_f:

$$-A_{el}(R^0) < E_F(C_1), E_F(C_f). \tag{11.1}$$

Assume moreover that R^- has still a positive electron affinity $A_{el}(R^-)$ and that it is sufficiently large to locate the energy level of R^{2-} at an intermediate energy between the Fermi energies of C_1 and C_f:

$$E_F(C_1) > -A_{el}(R^-) > E_F(C_f). \tag{11.2}$$

If conditions (11.1) and (11.2) are indeed satisfied,the following electron transfers (and only them) are allowed:

$$R^- + e^- \longrightarrow R^{2-} \qquad \text{at } C_1 \qquad\qquad \text{(I)}$$
$$R^{2-} \longrightarrow R^- + e^- \qquad \text{at } C_f \qquad\qquad \text{(II)}$$

Before discussing how this arrangement may be practically exploited, observe that a rotaxane with the above features can indeed be imagined. Suppose indeed that the ring is linked to a fullerene (that by itself is able to host 7 electrons) functionalized with a strong Brønsted acid AH (e.g., simply a carboxylic or sulfonic acid) whose acidity is neutralized via reaction with a strong Brønsted base BOH (e.g., –MgOH) covalently linked on side chain to axle A. After water elimination, the rotaxane will be formed by an axle hosting the covalently grafted, positively ionized, cation B^+ and a negatively ionized ring with strongly delocalized charge. The higher the delocalization, the lower the attractive electrostatic energy between A^- and B^+, and this energy may further be reduced by the presence of cryoptand (like oxo groups) on the axle in the vicinity of B^+.

To understand how such a complicated molecule can be exploited, assume that the ring, initially in charge state R^-, is in an intermediate position between C_1 and C_f and imagine one applies (by electrostatic coupling) a negative potential to C_f with respect to C_1. This potential will originate an electric field $\mathcal{E}$ eventually responsible for the drift of R^- to C_1. Once in the vicinity of C_1, the system will spontaneously equilibrate via reaction (I) (Fig. 11.6, top), and nothing else will happen until the voltage sign is reversed. When this occurs, the doubly charged ring, R^{2-} will drift to C_f and when in its vicinity the ring will transfer its second electron to the floating electrode according to reaction (II).

In so doing, the initial condition is restored thus making possible the repetitions of this cycle. The process is usefully repeated until the floating gate will attain a Fermi energy (under the specified voltage conditions) high enough to forbid reaction (II). In this way the machine behaves as an electron lift from C_1 to C_f.

If the process occurs only at a sufficiently strong electric field, while at weak fields ring R^- is electrostatically docked at its counterion, this machine can be used to charge floating gates in NVMs.

It is however noted that although the ring drift is most likely an efficient process (expectedly more efficient than the injection of hot electrons [346]), the continuous switches of the applied voltage are not only slow but are also responsible for large dissipation (it is sufficient to look at the cross section of Fig. 5.5 to realize the large resistive and capacitive parasitics involved in the crossbar structure). To make up for this difficulty, one can operate at fixed potential trying to exploit the Brownian motion.

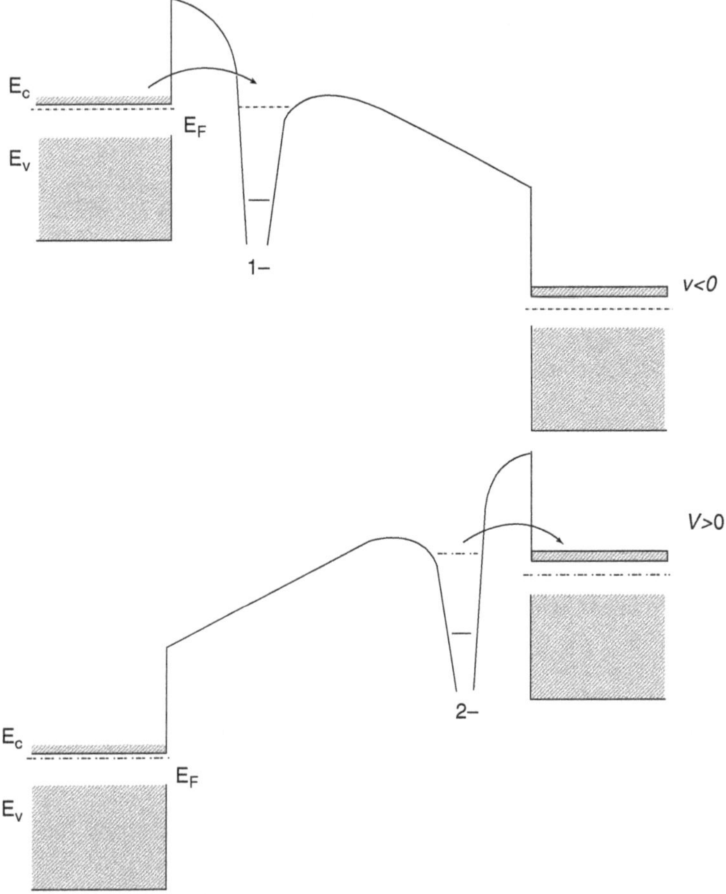

Fig. 11.6 Charge transfer from an electrode to a floating electrode via the alternate applications of electric fields of different orientations

11.3 Combining Ballistic and Brownian Motions

In this part it will be assumed that the ring can have two charge states, R^0 and R^- characterized by the following properties:

$$E_F(C_1) > -A_{el}(R^0) > E_F(C_f). \tag{11.3}$$

An idea for obtaining a redox center with the features of (11.3) is to insert covalently in the ring a transition metal M admitting two oxidation states $M^{(n)}$ and $M^{(n+1)}$, where M is covalently bonded to n and $n + 1$ atoms, respectively. When in oxidation state $(n + 1)$, the ring R is globally neutral and all valencies of M are fully saturated; in oxidation state (n) the ring is globally negatively charged and the unsaturated valencies of M are stabilized by suitable ligands possibly undergoing conformational changes after the redox reaction. A sketch of the considered situation is given in Fig. 11.5.

Assume that at time $t = 0$ the ring is neutral and in an intermediate position between C_1 and C_f. Between R is in the neutral state, it does not feel the electric field applied between the electrodes and is therefore subjected only to a Brownian motion due to its erratic collisions against the gas-phase molecules. The time required by the ring to dock either C_1 or C_f is of the order of $(\Delta x)^2/8D$, where Δx is the distance separating C_1 from C_f and D is the diffusion coefficient of the ring along the axle; let $\tau_{s\,1}$ and $\tau_{s\,f}$ be the sojourn times of R in close vicinity of electrodes C_1 and C_f, respectively (tentatively, $\tau_{s\,1} \approx \tau_{s\,f}$). Consider first the docking on C_f. The thermodynamics of the system inhibits any reaction so that the particle will leave C_f eventually diffusing to C_1 (after a time of the order of $(\Delta x)^2/2D$). Let $\tau_{0/-1}$ and $\tau_{-1/0}$ be the mean times required for the ionization and neutralization of the ring:

$$R + e^- \xrightarrow{\ \tau_{0/-1}\ } R^-, \qquad \text{at } C_1,$$
$$R^- \xrightarrow{\ \tau_{-1/0}\ } R + e^- \qquad \text{at } C_f.$$

Once in contact with C_1, if the ionization of R is a Poisson process, the probability p of its occurrence is given by

$$p = 1 - \exp\left(\tau_{s\,1}/\tau_{0/1}\right).$$

In the case there is no electron transfer from C_1 within the sojourn time, the system will evolve by a sequence involving the desorption of the neutral ring and its docking to the electrode until, after the repetition of p^{-1} cycles on the mean, the ring will undergo ionization. This ionization will produce a coupling of the ring with the applied electric field and its drift with average velocity $\bar{v}$ from C_1 to C_f. If the mean free path in the medium is smaller than Δx, $\bar{v}$ is controlled by the mobility μ in the medium: $\bar{v} = \mu \mathcal{E}$, with μ given by Einstein relation $\mu = (k_B T/e)D$. Once in contact with C_f, the ring will stay thereon (stabilized by the electric field) until it undergoes spontaneous neutralization (in a time $\tau_{-1/0}$) (Fig. 11.7).

Once the ring regains its neutral state, the initial condition is restored thus allowing the repetition of the cycle, as sketched in Fig. 11.8.

The cycle can be repeated until the increase of Fermi energy, produced by the accumulated electrons, thermodynamically disallow further electron transfer to the floating electrode. The electron transfer, indeed, increases the voltage (and thus Fermi energy) of the floating electrode by an amount e/C_{C_f}, where C_{C_f} is its capacity.

The described device is reminiscent of the ratchet-and-pawl machine sketched by Feynman in his famous *Lectures* [347], the pawl being provided by the ionization and by the applied electric field. Rather than trying to use this device for extracting energy from the medium, the device is operated as a nanoscopic Van de Graaff generator, where the drift of the electron carrier is provided by the applied electric field while its return to the low-voltage electrode is provided by Brownian motion.

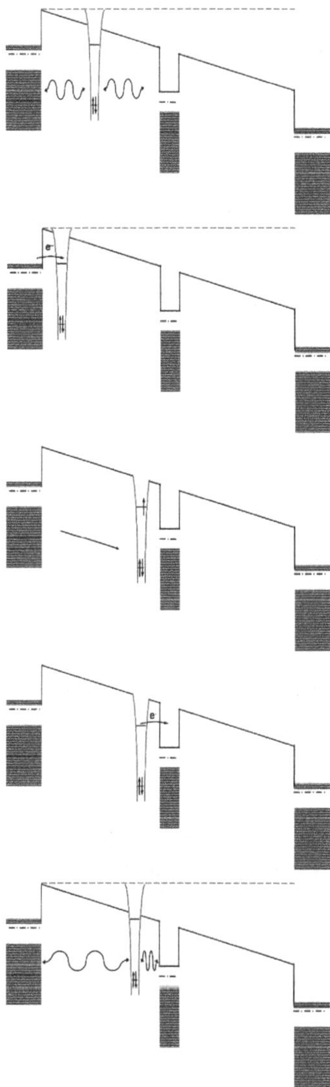

Fig. 11.7 Various states of the nano-Van de Graaff (from top to bottom): 1, the neutral ring in Brownian motion does not feel the applied electric field; 2, once in the vicinity of contact C_1, the ring undergoes ionization; 3, the ionized ring feels the applied electric field and shifts to electrode C_f; 4, the ionized ring is kept by the electric field in the vicinity of electrode C_f until it neutralizes ceding its extra electrode to the floating electrode; 5, the neutral ring does not feel any longer the applied field and is subjected to the Brownian motion only

What is of uppermost interest in this arrangement is the fact that many of the quantities that affect the observable increase of voltage of C_f are experimentally accessible (for instance: Δx, by choosing molecules with different axle lengths;

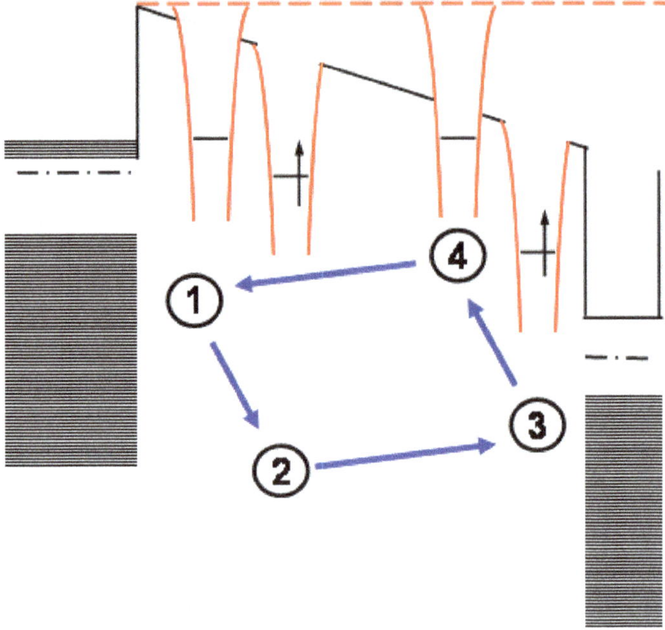

Fig. 11.8 The cycle allowing the rotaxane to behave as a van de Graaff generator

$\mathcal{E}$, by applying different voltages; D, by varying the gas pressure in the medium; etc.) thus allowing a study of the involved phenomena (electron transfer, Brownian motion, drift) at the scale of few, possibly one, molecules.

11.4 Brownian Motors

Two kinds of molecular motors have so far been considered:

Fully deterministic motor. The first device, described in Sect. 11.2, is fully deterministic, the alternate motion is imparted via the application of an alternate force by a macroscopic source, and friction reduces the device ability of transforming the external energy into work. The device may be viewed as a macroscopic motor shrunk to the molecular-mesoscopic scale.

Semideterministic motor. The second device, described in Sect. 11.3, is only partially deterministic because each working cycle is formed by two phases: in one phase the motion is deterministic and imparted by a source of (static) external force, whereas friction depresses the device ability of transforming external energy into work; in the other phase the motion is instead obtained exploiting the Brownian motion (the unavoidable counterpart of friction, because of the fluctuation-dissipation theorem [348]) of the mobile part of the molecule immersed in the dissipative medium.

As already discussed, the first device pays the advantage of its deterministic external control in terms of higher dissipation not only in friction but also in the control circuitry. Although this conclusion is based on the analysis of special devices, it seems to have a general validity.

The above classification does not exhaust all possible molecular motors and also a fully stochastic motor, where a directed motion is obtained exploiting the fluctuation of the dissipative medium even in the absence of any external source of force, deserves consideration. Of course, extracting mechanical work from a system at assigned temperature is inconsistent with the Second Law of Thermodynamics[3] so that such Brownian motor may operate only via the consumption of an external free energy:

Fully stochastic motor. In this device a directed motion is obtained exploiting the fluctuation of the dissipative medium, in the absence of any external source of force, via the consumption of free energy.

A lot of work has been devoted to the understanding of the detailed behavior of biological motors.

A highly simplified model starts from the analysis of the diffusion along a chain. Let the chain be formed of n equivalent, linearly arranged, units. Each unit is adapted to bind specifically to a certain species (molecule or ion) M.

The equilibrium configuration of M in each unit is centered on a sharp minimum, but the thermal reservoir at finite temperature allows the diffusion from one site to the adjacent ones. The rate of thermally assisted diffusion is given by $v_0 \exp(-\epsilon/k_B T)$, where v_0 is the vibration frequency along the chain and ϵ, the activation energy for migration, is the mean energy required to overcome the barrier separating the minima. If they are equivalent, the energy barriers for backward and forward migration are the same, so that in the thermal reservoir a particle initially localized on a minimum will experience a Brownian motion where forward and backward diffusions occur with the same probability, with a diffusion coefficient D given by

$$D = v_0 \lambda_0^2 \exp(-\epsilon/k_B T), \tag{11.4}$$

where λ_0 is the distance between neighboring minima.

A preferential diffusion (say forward) is possible superimposing on the potential generated by the linear chain a (generally weak) external potential U, whose gradient is responsible for a drift of M. The drift velocity v is given by

$$\mathbf{v} = -\mu \nabla U$$

---

[3] Fluctuations may be viewed as local and temporary violations of the Second Law; the space and time limits of such violations are discussed in [349].

where μ is the mobility. The external force sustains a velocity, rather than an acceleration, because the Brownian motion drags the ballistic motion; this is contained in Einstein's expression of the mobility:

$$\mu = k_B TD.$$

The existence of an external field is not the only case allowing the motion in a preferential verse. An example is obtained by considering a linear chain generated potential like that sketched in Fig. 11.9, where the total binding energy of M in

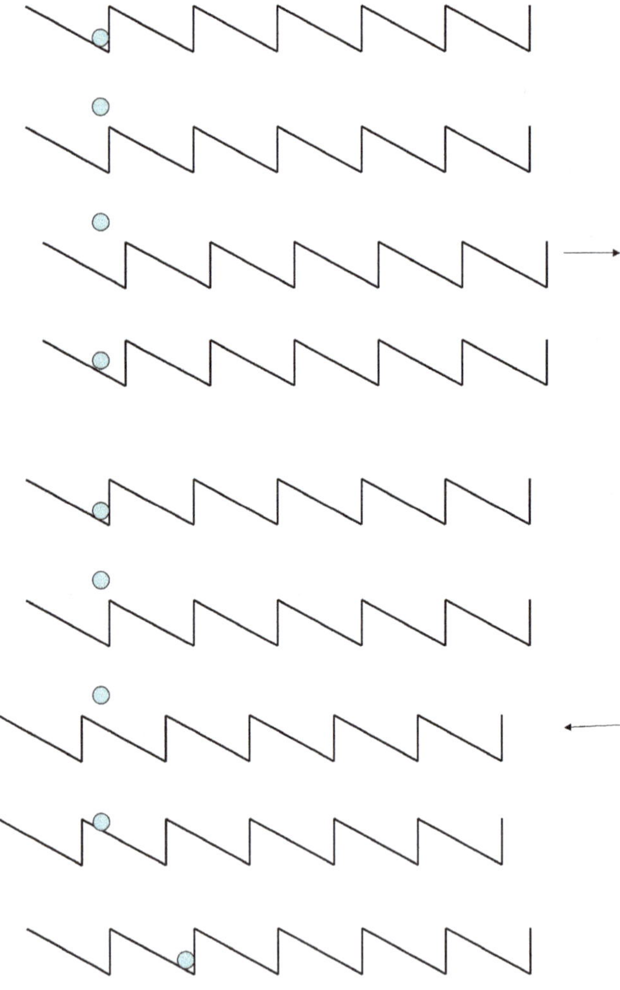

Fig. 11.9 How thermal excitation of an adsorbed molecule and the vibrations of the adsorbent (generating a suitably shaped potential) can combine to produce biased diffusion

each unit has a strongly asymmetric landscape, varying spatially from the relative minimum very weakly in a given verse ("backward," say toward left) and very fast in the opposite verse ("forward," toward right).

Of course, (11.4) holds true irrespective of the energy landscape. However, M can attain a sufficient energy to overcome the energy barrier separating two units of the chain only because it is embedded in a thermal reservoir. Since the cause that allows M to acquire energy (the thermal reservoir) is responsible for energy fluctuations also of the chain, ϵ must be considered as a kind of effective energy for the strongly coupled system. The potential generated by the chain must therefore be viewed as fluctuating due to thermal vibrations.

Imagine now that the chain is subjected to further vibrations (with consumption of energy) in addition to the ones imparted by the thermal reservoir. The major effect of the thermal reservoir is to bring the particle (with energy around ϵ) from the minimum to a configuration close to the top of the energy barrier separating the units. The additional vibrations may have two effects: in approximately 50% of cases M will be forced in the region of the weak potential whose effect is to drive M forward, in the nearby minimum; in the remaining cases M will be swiftly brought to the original position. Thus, the combined effects of the thermal reservoir and additional vibration are to move M forward or to confirm it in the original position. The repetition of these events will eventually result in a biased (forward) diffusion, even against concentration gradients.

It is stressed that such motions exploit the Brownian motion (to allow the diffusing species to reach the top of the energy barrier) but is possible only in the presence of an additional motion sustained by the dissipation of energy.

Biological motors are essentially based on the above principle, although the motion involves complicated conformational changes fueled by the hydrolysis of ATP to adenosine diphosphate (ADP) [350–352].

Of course, the above considerations are addressed neither to describe any biological motor nor to account for their functioning: The complicated interplay of chemistry, biology and physics at the bases of biological motors is discussed in several articles (see, for instance, [350–352]); rigorous mathematical treatments are also available (see, for instance, [353–356]); and synthetic examples of such motors are now well known [336, 357, 358].

Chapter 12
Nanobiosensing

Most physicists will agree with the following words (taken from Lebowitz [359]):

> Nature has a hierarchical structure, with time, length, and energy scales ranging from the submicroscopic to the supergalactic. Surprisingly, it is possible, and in many cases essential, to discuss these levels independently – quarks are irrelevant for understanding protein folding and atoms are a distraction when studying ocean currents. Nevertheless, it is a central lesson of science, very successful in the past 300 years, that there are no new fundamental laws, only new phenomena, as one goes up the hierarchy. Thus arrows of explanations between different levels always point from smaller to larger scales, although the origin of higher-level phenomena in the more fundamental lower-level laws is often very far from transparent.

To support this statement, it suffices to note that the properties of thermodynamic systems are described (with thermodynamics) in terms of their constituting phases, supposedly given; the properties of phases are described (with statistical mechanics) in terms of their constituting molecules, supposedly given; and the properties of molecules are accurately described (with quantum mechanics) in terms of their constituting electrons and nuclei, supposedly given. Figure 12.1 sketches how ordinary matter can be hierarchically organized in strata and described in a reductionistic manner.

Together with quantum mechanics, molecular biology is the scientific discipline of major success in the twentieth century. Molecular biology may be seen as the extreme attempt to reduce the properties of living systems to those of their molecular constituents.

12.1 Reducing Cell Biology to Molecular Biology

In extreme synthesis, molecular biology summarizes life phenomena in terms of its Central Dogma: life is controlled by genetic properties encoded in a family of molecules (DNAs, deoxyribonucleic acids) able:

- To replicate themselves
- To transcript the code in auxiliary molecules (RNAs, ribonucleic acids)
- To translate them into functional units (proteins)

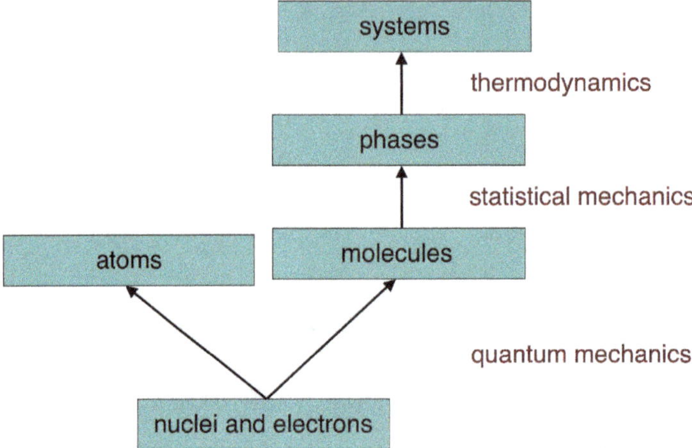

Fig. 12.1 The hierarchical organization of ordinary matter and its reductive descriptions

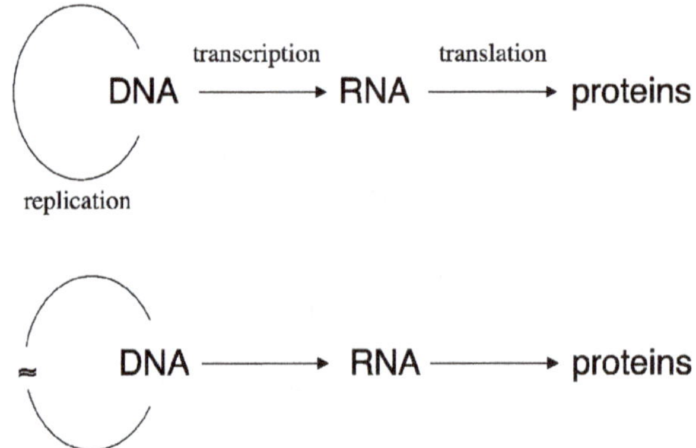

Fig. 12.2 The expressions of the genetic codes (*top*), stressing the fact that the duplication of DNA is subjected to errors (*bottom*)

Figure 12.2 (*top*) sketches these functions.

DNA replication, however, is a near equilibrium process and is thus subjected to errors. The consideration of the degree of randomness associated with replication and eventually responsible for mutations leads eventually to the world (so well described by Monod [360]) dominated by chance and necessity. Figure 12.2 (*bottom*) stresses the possibility of errors.

Even endowing the scheme of Fig. 12.2 (*bottom*) with the necessary feedbacks (for which proteins control duplication, transcription, and translation; see

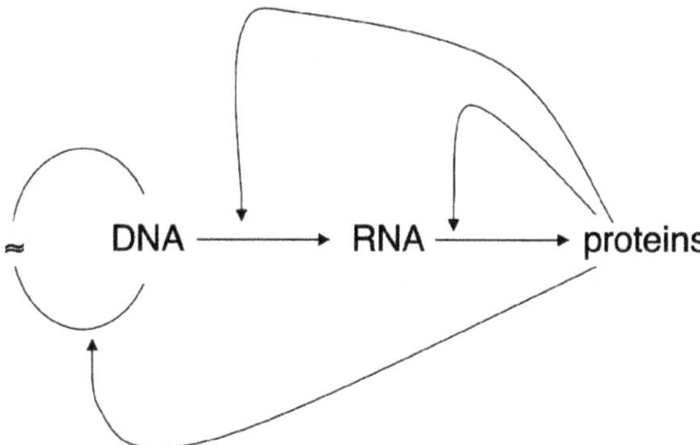

Fig. 12.3 Specifying the feedbacks in the Central Dogma of molecular biology

Fig. 12.3),[1] molecular biology remains a reductive theory, for which the properties of the biological systems are nothing but average properties of the constituting molecules (although strongly interacting).

Once applied to the description of the cell, however, the above view appears however too rigid because of the combination of elemental, molecular, and phase heterogeneities with chaotic behavior.

Elemental Heterogeneity Although only a score of elements are necessary (in ponderable amount or as traces) for the life of the cell, any cell contains not only those elements, but also practically all other elements in the environment to which the cell has been exposed and at concentrations depending on ambient concentrations.

Molecular Heterogeneity The functional properties of proteins are controlled by their conformations. Even the simplest protein (myoglobin) appears to be formed by a heterogeneous collection of the same molecule in different conformations.

Phase Heterogeneity Although the folding properties of proteins go beyond current computability,[2] there is wide agreement on the fact that the conformations of proteins are not, in general, the equilibrium ones and that they are controlled by the growth conditions in the aqueous phase where the synthesis has been carried out. In turn, the structure of the medium is determined by the interactions (of

---

[1] Another pathway, where the duplication of DNA is controlled by RNA (*inverse transcription*), is not shown. This pathway is employed for reproduction by viruses containing only RNA as genetic material (retroviruses).

[2] In 1969, Levinthal noted that, because of the very large number of degrees of freedom in an unfolded polypeptide chain, the molecule has an astronomical number ($\sim 3^{300}$) of possible conformations. If the protein is to attain its correctly folded configuration by sequentially sampling all the possible conformations, it would require a time longer than the age of the universe to arrive at its correct native conformation (*Levinthal paradox*) [361]. This is true, even if conformations are sampled at rapid (nanosecond or picosecond) rates. A peptide consisting of just five amino acids can fold into over 100 billion possible structures.

the 10^3–10^4 mutually interacting, different proteins contained in the cell), with the medium dominated by the heterogeneity of backbone and side-chain polar groups and characterized by allosteric transition and hysteresis effects [362].

Chaotic Behavior The same substance may participate in different reaction pathways and, conversely, many other substances not participating in those pathways may influence them. Generally, these interactions are nonlinear so that the evolution of the system can be predicted only at coarse grain (qualitative description) or not be predicted at all.

The consideration of all these factors suggests that both the *direct problem* of predicting on microscopic levels the behavior of the cell, and the *inverse problem* of detecting pathologic deviations from the physiological one, are beyond the current computational capabilities.

It seems natural therefore, to bypass the attempt at accounting for the existence and functions of the cell in terms of its constituting molecular parts and to consider life as an emergent property. Systems biology is the branch of biology addressed to the systematic study of complex interactions in biological systems, in the perspective of integration rather than of reduction.[3]

12.2   From Molecular Biology to Systems Biology

The emergent feature that can be assumed to define a living system, more than any other, is its metabolism. Any cellular system is associated with an oscillating *metabolic field* with quasi-steady frequencies, sustaining the transport of anabolites from the environment and of catabolites in the opposite verse.

Starting from this elementary observation, Rashevsky and his school developed a theory, summarized in the book *Mathematical Biophysics* [308] (a masterpiece of the scientific thought of the twentieth century) where even very complex properties, like nerve excitability, are derived from the properties of the diffusion field and of its confinement in a restricted space.

Rashevsky's theory, however, was well developed before the epochal discoveries of molecular biology (protein structure and function, Central Dogma of biology, molecular genetics, etc.) whose mainstream has, conversely, dealt only marginally with the characters of Rashevsky's theory.

In recent years, however, metabolism has returned to play a central role in biology. That metabolic diseases at the organism level often have familiar bases and may be reduced to genetic disorder is too well known to deserve discussion. That

---

[3] "Particularly from year 2000 onwards, the term is used widely in the biosciences, and in a variety of contexts. Because the scientific method has been used primarily toward reductionism, one of the goals of systems biology is to discover new emergent properties that may arise from the systemic view used by this discipline in order to understand better the entirety of processes that happen in a biological system." [363]

a metabolic pattern at the lowest functional level (the cellular one) may be used as a tool to sense genetic disorder (eventually culminating in cancer) is, on the other hand, not so trivial, and the opinion that information at subcellular level may be used for preventative and curative medicine is gaining increasing consensus [364, 365].

As metabolism is considered an emergent property, its description cannot be reduced in terms of the underlying molecular properties only, but rather must be assigned a priori on empiric or speculative bases.

In this way, the theory will contain free parameters describing the *internal state* of the living system; these parameters λ_i $(i = 1, \ldots, n)$ will be in mutual relation through a set of Γ equations that also specify the interactions of the cell with the external world, characterized by the thermodynamic variables x_I $(I = 1, \ldots, N)$:

$$\Phi_\gamma(\lambda_1, \ldots, \lambda_n; x_1, \ldots, x_N), \quad \gamma = 1, \ldots, \Gamma.$$

Any theory of cells will thus be characterized by the set $\{\{x_I\}, \{\lambda_i\}, \{\Phi_\gamma\}\}$.

The objective of systems biology is the identification of the best choice of $\{\{x_I\}, \{\lambda_i\}, \{\Phi_\gamma\}\}$. In view of the inherent complexity of the ab initio (bottom-up) description of cell behavior, all cell models are wrong, and some more wrong than others. The selection of less wrong, potentially useful, models can be done in a top-down fashion sensing the cellular behavior and selecting (essentially via genetic algorithms) the simplest model accounting for observed behavior within an assigned accuracy.

Clearly enough, increasing the level of sensing accuracy (finer space localization, greater number of parameters, etc.) will require a better description, so that one can image a roadmap where the model is improved starting from a few parameters and poor space resolution, and progressively increasing the spatial resolution and the number of parameters.

12.3 Sensing as a Key Tool for Systems Biology

Without pretending to assign any priority, one of the first examples where the high spatial resolution offered by the IC technology could be used for solving biological problems was perhaps proposed in [366]: Starting from the observation that when an α particle strikes a memory cell of a DRAM, it produces the recombination of the metastable charge packet stored therein (single event upset), whereas thermal neutrons have a negligible interaction with the silicon, Cerofolini et al. suggested that information on boron or lithium distributions inside cells could be obtained by simply depositing the cell onto a DRAM, exposing the resulting structure to a stream of thermal neutrons, activating the following reactions:

$$^{7}\text{Li} + {}^{1}\text{n} \longrightarrow 2\alpha,$$
$$^{10}\text{B} + {}^{1}\text{n} \longrightarrow {}^{7}\text{Li} + \alpha,$$

Fig. 12.4 Microphotograph of an amoeba covering approximately 40 cells of a DRAM

and detecting which cells give single event upsets. Figure 12.4 shows that this kind of radiography was able to locate (when the method was proposed, in the early 1980s) the considered elements to within $(1/40)$th of the amoeba.

Much more interesting is certainly the hybrid combination of neurons with nanoelectronic circuitry. The major problems here come from the fact that mature neurons survive for very short time in vitro (or in silico) so that most of the work has been carried out with partially undifferentiated neuroblastomas (see [367] for an early work in this field). Once one has learnt how to allow the neuron to survive in silico, the sensing of its membrane potential with a nanoscopic crossbar (functionalized to allow the adhesion and to be sensitive to the ion distribution on the neuron membrane) will succeed in its determination with a resolution on a length scale coinciding approximately with the nearest neighbor distance between the ion channels that determine cell excitability.

12.4 From ICs to Nanobiosensors

Subcellular sensing, however, requires nanoscopic circuits with complexity exceedingly larger than that of microprocessors; technology for subcellular sensing is thus motivated by a huge market only. The health market satisfies this need; nonetheless, the technology for subcellular sensing has a higher chance of success if it can exploit the progress of the IC technology.

12.4.1 The Incremental Increase of Complexity of ICs and Sensors

On the other side, it is tempting to take advantage of the increasing complexity allowed by the IC technology to probe with greater detail (more parameters on shorter length and time scales).

Consider the following classification[4]:

Moore denotes the past and current IC technology.

More Moore indicates the near-future electronic devices and circuits of the near future, hypothesized in the Roadmap. Scaling electronic devices and circuits will continue into the next decade; it will be driven by Moore's laws, following the technologies described in the ITRS with CMOS, the main semiconductor technology platform.

More than Moore summarizes the extension of current CMOS processes to add new functionalities. Current CMOS processes can be used to develop new micro- and nanodevices with extended functionalities, such as sensors, MEMSs, or nanoelectromechanical systems (NEMSs). More-than-Moore technologies include heterogeneous integration, advanced packaging, three-dimensional integration, biomedical electronics, autonomous embedded appliances, and photovoltaics.

The above frames can be arranged in the diagram on the left of Fig. 12.5, where even in Moore and More Moore there is an increase of complexity, in addition to the increase of density. The complexity, however, is limited to electrical functions other than the digital ones: namely, the management of power (smart power circuits), the transmission of data (radio-frequency circuits), etc.

However, whereas the CMOS technology is characterized by the need for insulating the FETs and their interconnects from the environment (so that the FETs are separated from the environment by a number of layers with the total thickness on the micrometer length scale), the sensor technology is characterized by the need of putting the sensitive element in contact with the environment (see Fig. 12.6). Because of these opposite needs, integrating logic with sensing is obviously difficult.

The difficulties of integrating new functions into a certain technology are the cause for the delay in the development of high-density smart sensors. The time evolution of the technology is thus the one sketched on the right of Fig. 12.5.

---

[4] The occurrence of the name of Moore in the following may seem excessive, and it would certainly be so if the reason were simply the statements of Chap. 2. Moore, however, has other merits: he developed the silicon-gate technology [25] (invented by Faggin and coworkers [23, 24] on the industrial scale was a coinventor, with Hoff, of the microprocessor [368]) and a cofounder, with Noyce, of Intel.

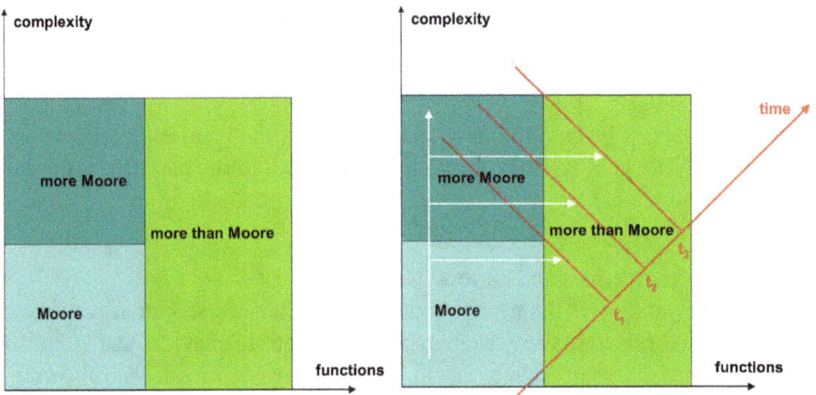

Fig. 12.5 Classification (*left*) and time evolution (*right*) of ICs according to complexity and functions

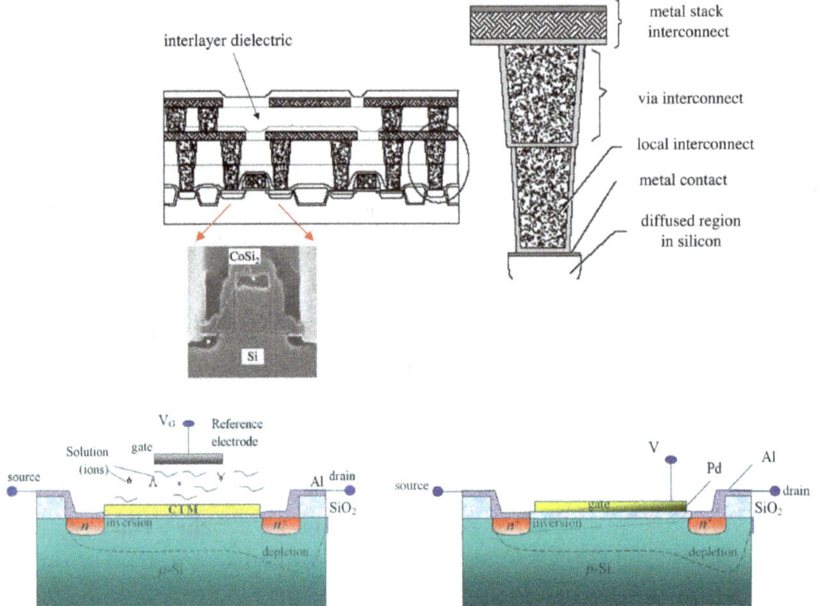

Fig. 12.6 Comparing the structure of a real IC (*top*) with those of ion-sensitive (*bottom and left*) and gas-sensitive (*bottom and right*) FETs

12.4.2 The Shift of Paradigm

As discussed in Chap. 2, sooner or later it will be economically impossible to overcome the limits of photolithography. Preparing ICs with TSI complexity will only be possible employing methods like those sketched in the first part of this book, thus going beyond Moore:

Beyond Moore sketches a future where CMOS technology will survive in hybrid combination with genuine nanotechnologies, possibly based on new devices (quantum dots, reprogrammable molecules, etc.), architectures (crossbar), or principles (quantum computing).

Adding new functions to circuits with TSI complexity produced with nonlithographic methods will eventually bring the technology outside the walls – *extra muros*:

Extra muros hypothesizes a world where nanodevices with new functions are integrated in hybrid combinations with post-CMOS circuits for the full exploitation of their nanoscopic properties.

The positioning of circuits beyond Moore and *extra muros* in the {complexity, density} plane is sketched in the left part of Fig. 12.7. The right part of this figure shows the expected time evolution.

Nanobiosensing is expected to be the final step of a gradual evolution eventually leading (from current ICs) to circuits of TSI complexity with many functions, and spatial resolution sufficient to sense living systems at deep subcellular level, embedded in silicon-based logic circuit.

Crossbars with crosspoints allowing a resolution of 20–30 nm are currently achievable with the techniques described in Chap. 5. The crossbar is thus a candidate for subcellular biosensing, provided one is able to functionalize the crosspoints and link them to the external world.[5]

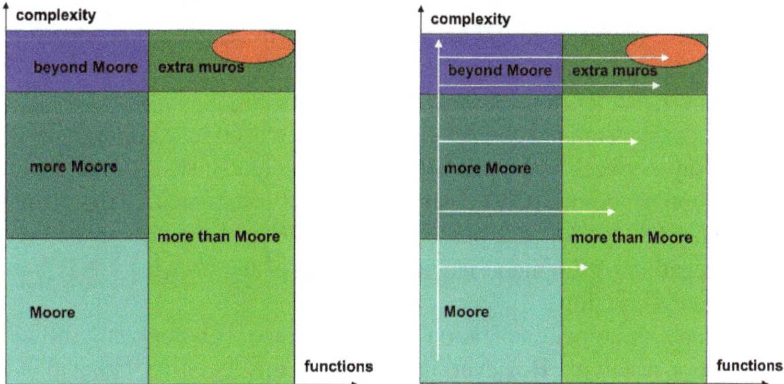

Fig. 12.7 Classification (*left*) and time evolution (*right*) of conventional (*lower boxes*) and new (*upper boxes*) devices according to complexity and functions (the *oval* denotes the region of nanobiosensors)

---

[5] A review of the current status of bionanopatterning of surfaces is given in [369].

12.5   A Roadmap for Nanobiosensing

Since the crossbar structure can be prepared in a stage close to the back-end (see Chap. 4), the overall fabrication process can be arranged in such a way as to have the crossbar near the region to be sensed, without destroying the nearby hosting CMOS circuitry. The combination of this fundamental feature with the spatial resolution consistent with its preparation techniques (Chap. 5) makes the crossbar structure a candidate for nanobiosensing.

12.5.1   Nanobiosensing In Vitro

Nanobiosensors for the subcellular analysis in vitro are already possible by exploiting the potentials of the crossbar.

One can imagine that, after the definition of a bottom array of p-doped poly-Si wires (Fig. 12.8a), a thin film of material displaying highly nonlinear conduction (SiO_2, with conduction controlled by the electric field at the poly-Si surface, being a plausible candidate) is deposited (Fig. 12.8b), and another array of n-doped poly-Si wires is defined on it (Fig. 12.8c). After that, the whole structure is covered by a relatively thick insulator (Fig. 12.8d) that is etched using a mask with the same pattern as in the top electrode, but oriented perpendicularly. This will result in the exposure of n^2 poly-Si zones, doped n-type and approximately square in shape (Fig. 12.8e).

The immersion of the resulting structure in the salt solution of a reducible metal (like nickel, copper, etc.) will result in its electroless deposition in elemental form on the exposed silicon. The deposited film may be thermodynamically or kinetically controlled to have an assigned thickness. A thermal treatment of the structure will result in the formation of one or more islands on the silicon, with size controlled by surface and interfacial tension, as shown in the top of Fig. 12.9.

After that, the system is exposed to an ethylene or acetylene atmosphere at high temperature (say, between 700 and 1,000°C), and the metal islands catalyze the formation of carbon nanotubes (CNTs) whose diameter is assigned by the diameter of the metal catalyst.[6] The process proceeds, as sketched at the bottom of Fig. 12.9, with the segregation of the metal island at the top of the growing CNT, thus preserving the catalytic action.

The last stage, the functionalization of the nanotube, is notoriously difficult because of its poor reactivity. However, in addition to the chemistry targeted at direct functionalization, the metal cluster at the top of the CNT can be exploited as a Trojan horse. Derivatizing the CNT with crown ethers [341], as sketched in Fig. 12.10, will

---

[6] With carbon nanotube it is denoted an all-carbon hollow graphitic material with a high aspect ratio. Formally a CNT is obtained rolling up single or multiple graphene sheets to give a single-walled CNT (SWCNT) or a coaxial multiple-walled CNT (MWCNT), respectively. The CNT length is typically in the interval 10^2–10^4 nm and diameter 0.2–4 nm for SWCNTs or 2–100 nm for coaxial multiple-walled MWCNTs [370].

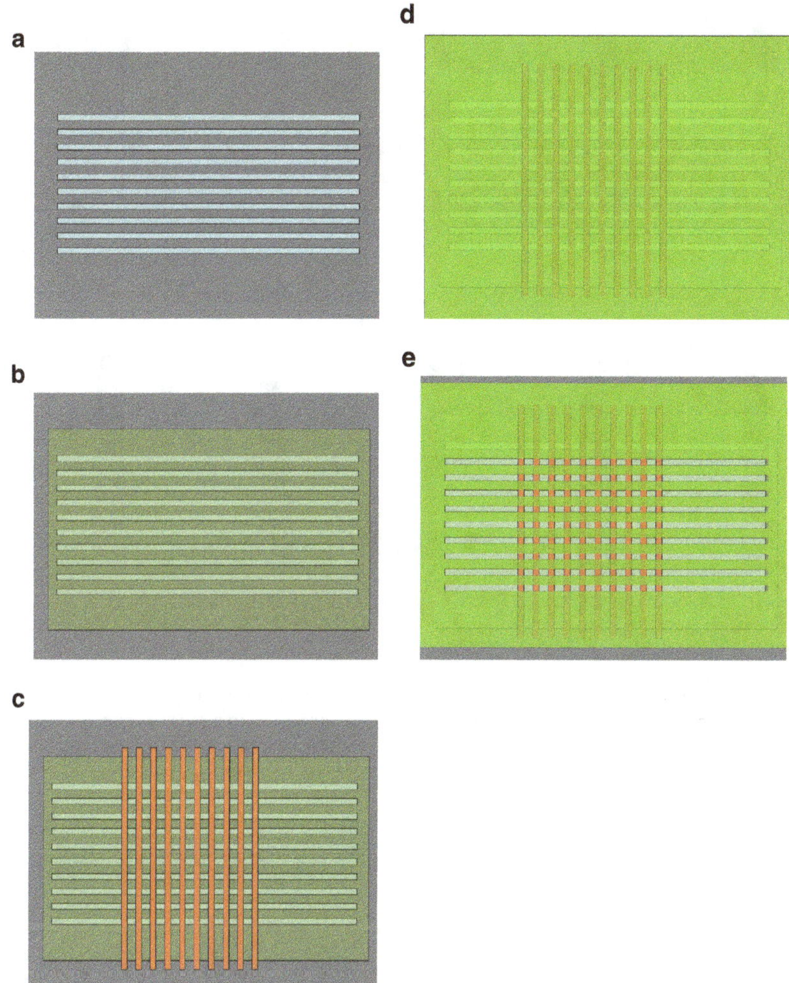

Fig. 12.8 The process for the preparation of a crossbar suitable for further functionalization: (**a**) definition of the bottom array of poly-Si wires; (**b**) deposition of a material with strongly nonlinear conductance; (**c**) definition of the top array of poly-Si wires; (**d**) deposition of an insulating film; and (**e**) etching of the outer insulating layer with a mask for the exposure of a matrix of n^2 squares on the top wire array

result in sites able to capture alkali ions[7] in a highly selective way; grafting the CNT with calixarenes [341] modified to contain an electrophilic center (or even a Lewis acid) will result in sites able to capture anions with selectivity controlled by the calix shape and size.

---

[7] The selectivity is determined by the crown ether: 18-crown-6 and 15-crown-5 are highly selective for K^+ and Na^+, respectively.

Fig. 12.9 Preparing the CNT matrix by the deposition of metal islands on the exposed n-type silicon (*top*) and the metal-catalyzed growth of CNTs by exposure to a hydrocarbon atmosphere at high temperature (*bottom*)

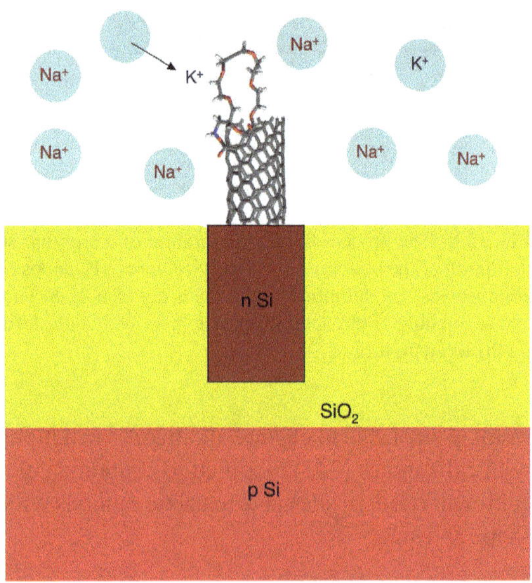

Fig. 12.10 How CNT functionalized with crown ethers may be used for the chemically selective and space-resolved detection of alkali ions

This process completes the sensor: the (highly selective) capture of an ion by the functionalized CNT will produce a significant change in the voltage distribution in the underlying crosspoint, which can be detected by measuring the current flowing through it because of the highly nonlinear $I-V$ characteristic.

12.5.2 Nanobiosensing In Vivo

Nanobiosensors are the ultimate frontier of implantable biosensors. Although they may appear quite distant by current possibilities, trying to design a roadmap showing a path for their production is not merely science fiction, but rather, may be useful for identifying bottlenecks and the ways to avoid them. Walking along such a road will certainly be very difficult; a sense of that path, however, is conferred by the fact that the hypothesized subcellular sensors result from a combination of solutions, for which a demonstration of feasibility has already been given.

In vivo nanobiosensors must have additional features with respect to those operating in vitro. In particular, in addition to its sensing ability, the implantable nanobiosensor must be fueled to be mobile (to float to the target cells), and have a minimum of intelligence (to direct itself to the target cells and to manage the information resulting from sensing). Above all, all this activity makes no sense unless the data are eventually transmitted to a human observer.

Satisfying all these functions is not simple. Observe, however, that an area equivalent to that of sensing (say $10^2\,\mu m^2$) can host approximately 10^5 crosspoints, consistent with a logic of LSI–VLSI complexity. More difficult is imparting mobility to the nanobiosensor. However, hybrid solutions (where the motion is imparted by the derivatization of the sensor with biomotors) can be hypothesized: for instance, Montemagno and Bachand have reported the construction of nanomechanical devices powered by biomolecular motors [371], Kim and Breuer have described the successful use of live bacteria as mechanical actuators in microfabricated fluid systems [372], and Behkam and Sitti have exploited bacterial flagella for propulsion and motion control of microscale objects [373]. An enormous advantage of this solution is the fact that the motion does not require an alien input of energy – for this, the chemical energy available in the organism (in the form of ATP) can be exploited. The same form of energy can also be exploited for powering electrical circuitry, because biomotors are reversible and can operate as engines too. It is also to be noted that the back of the device may be exploited for the preparation of a supercapacitor as a buffer element for power supply.

To avoid the immune response of the organism, the nanobiosensor must be coated with a film not felt as hostile by the organism. That this is within reach of current technology is evident from the several demonstrations of lipid monolayers or bilayers (mimicking biological membranes) supported on solid or polymer surfaces on even large areas [374]. Of course, the supported films are only weakly bound and are thus poorly stable; a larger stability, however, can be achieved by bonding molecules covalently (e.g., via silanization) with carboxylic terminations, mimicking the outer surface of cells to the sensor surface.

A seemingly unsurmountable barrier is represented by data transmission: radio-frequency transmission requires an antenna whose size is most likely in the millimeter length scale. However, one can hypothesize that the sensor is a part of a fleet with its basis on an implanted and immobile circuit, with area of the order of $1\,cm^2$, and thus able to host GSI logic and to manage the transmission of relevant data to the human observer.

At present, it is difficult to imagine whether current technologies will allow the limits of sensing to be overcome, to enter the world of nanosurgery. However, following the announcement on December 2002 by the U.S. National Institutes of Health (NIH) of a 4-year program for nanoscience and nanotechnology in medicine, a Roadmap to Nanomedicine has been sketched [375]:

> Burgeoning interest in the medical applications of nanotechnology has led to the emergence of a new field called nanomedicine. Most broadly, nanomedicine is the process of diagnosing, treating, and preventing disease and traumatic injury, of relieving pain, and of preserving and improving human health, using molecular tools and molecular knowledge of the human body. The NIH Roadmap's new Nanomedicine Initiatives, first released in late 2003, 'envision that this cutting-edge area of research will begin yielding medical benefits as early as 10 years from now' and will begin with 'establishing a handful of Nanomedicine Centers ... staffed by a highly interdisciplinary scientific crew including biologists, physicians, mathematicians, engineers and computer scientists gathering extensive information about how molecular machines are built' who will also develop 'a new kind of vocabulary – lexicon – to define biological parts and processes in engineering terms.' In the relatively near term, over the next 5 years, nanomedicine can address many important medical problems by using nanoscale-structured materials and simple nanodevices that can be manufactured today. This includes the interaction of nanostructured materials with biological systems. Over the next 5–10 years, biotechnology will make possible even more remarkable advances in molecular medicine and biobotics – microbiological robots or engineered organisms. In the longer term, perhaps 10–20 years from today, the earliest molecular machine systems and nanorobots may join the medical armamentarium.

Chapter 13
Abstract Technology

The unit operations characterizing planar technology have been briefly mentioned in Sect. 4.2. Of course, the accurate description of processes like the deposition of poly-Si or Si_3N_4, the oxidation of silicon, the plasma or sputter etching of Si, SiO_2, Si_3N_4, metals, etc., requires a detailed knowledge of the reaction kinetics and the solution of the diffusion–reaction equations with moving boundary conditions, coupled with the rheological equations of each reactant and product.

Designing a process architecture, however, does not require of all the above, and to some extent one may deal, for instance, with poly-Si or Si_3N_4 depositions as representative of a unique process (conformal deposition), the difference being given by the constituting materials only (poly-Si instead of Si_3N_4, in the considered example).

Most of the unit operations considered in this book may ultimately be reduced to the following:

- *Conformal deposition*
- *Isotropic etching*
- *Selective etching*
- *Directional etching*
- *Directional deposition*

in addition to the one at the basis of microelectronics,

- *Lithographic definition*

Each of these can be thought of as a transformation that operates on assigned bodies transforming them into other bodies.

This chapter is intended to build the *Elements of Abstract Technology* in the same spirit as Euclid's *Elements of Geometry*, using the said operations instead of the construction with straightedge and compass. The use of highly idealized unit operations (instead of their practical realizations) is, indeed, reminiscent of the use of ideal straightedge and compass (instead of their practical realizations).

It is anticipated that the operations of Abstract Technology are neither the ones of Euclidean geometry (they are neither translations nor rotations) nor those of topology. To be convinced of this statement, let us think simply of lithographic definition: in general, this process transforms a simply connected layer of a given material into a set of complementary nonconnected or multiply connected regions of two different materials.

Of course, there are superhighways of thought (geometry, topology, chemistry) for describing single fragments of the planar technology, but no general theory for all of them. From another view point, paraphrasing Hamlet [376], there are more phenomena on and at surfaces than are dealt with in any current theory. This chapter is addressed to formulate a theoretical framework for the description of the changes undergone by any body when subjected to a few unit operations characteristic of planar technology.

13.1 Material Bodies and Surfaces

The mathematical description of shape is a notoriously difficult matter [377]. Describing how a body changes its shape according to physical chemical processes, more complex than simple translation or rotation, is even more complicated and requires a specification of the material nature of the body. The minimal characterization of a body is in terms of its geometry and constituting materials.

Definition 1 (Simple body) A simple body $\mathcal{B}_X$ is a three-dimensional closed connected subset B of the Euclidean space $\mathbb{E}^3$ filled with a material X:

$$\mathcal{B}_X = (B, X).$$

What a material is, here, a primitive concept. It may be a substance, a mixture, a solution or an alloy. The emphasis is on homogeneity – whichever region, however small, of B is considered, the constituting material is the same. When not necessary, the index denoting the constituting material will not be specified.

In view of the granular nature of matter, considering the limit for vanishing size does not make sense. Here and in the following, when dealing with the concept of vanishing length (as implicitly happens when considering the frontier B^* of B), we mean that the property holds true on the *ultimate length scale*. Although the ultimate length scale is an operative concept related to the probe used for observing the body, we have in mind, the molecular one. Analogously, we consider a body as indefinitely extended when its size is much bigger than the size variations induced by the considered processes. In this sense, a *wafer* is a simple body indefinitely extended over two dimensions.

Definition 2 (Composite body) Consider the N-tuple of simple body ($\mathcal{B}_{X1}$, $\mathcal{B}_{X2}$, ..., $\mathcal{B}_{XN}$) formed by nonoverlapping sets ($B_1, B_2, \ldots, B_N$), with

$$\forall I, J \left(B_i \cap B_j = B_i^* \cap B_j^* \right),$$

and let

$$B = \bigcup_{i=1}^{N} B_i.$$

If B is connected and the simple bodies $\mathcal{B}_{X^1}, \mathcal{B}_{X^2}, \ldots, \mathcal{B}_{X^N}$ contain at least two different materials X^I and X^J, then the N-tuple is a composite body.

Definition 3 (Interface) Let a composite body $\mathcal{B}$ be formed by simple bodies $\mathcal{B}_{X^1}$ and $\mathcal{B}_{X^2}$ with $X^1 \neq X^2$; the region

$$F_{1|2} = B_1^* \cap B_2^*$$

defines the interface between $\mathcal{B}_{X^1}$ and $\mathcal{B}_{X^2}$.

For any $\mathbf{x}$ in $F_{1|2}$, all neighborhoods, however small, contain materials X^1 and X^2.

Definition 4 (Total surface) The set

$$S_{\text{tot}} = B^*$$

is the total surface of $\mathcal{B}$.

Proposition 1 *Let a composite body $\mathcal{B}$ be formed by the N-tuple of simple body* $(\mathcal{B}_{X^1}, \mathcal{B}_{X^2}, \ldots, \mathcal{B}_{X^N})$. *Then,*

$$S_{\text{tot}} = \bigcup_i B_i^* \setminus \bigcup_{j,k} F_{j|k}$$

Definition 5 (Surface) Let $\mathcal{B}$ be a composite body, and $B_\bullet$ be the smallest simply connected (i.e., without holes) set containing B. Then,

$$S = B_\bullet^*$$

is the outer surface (or simply surface) of $\mathcal{B}$.

Definition 6 (Inner surface) Let $\mathcal{B}$ be a composite body. Then,

$$S_{\text{in}} = S_{\text{tot}} \setminus S$$

is the inner surface of $\mathcal{B}$.

13.2 Processes Controlled by Geometry

The processes considered in this section are controlled only by the geometric and topologic properties of the body.

13.2.1 Conformal Processes

Definition 7 (Delta coverage) For any body $\mathcal{B}$ with surface S, the additive delta coverage $\overline{D}_{+\delta}$ of thickness δ is the set

$$\overline{D}_{+\delta} = \{\mathbf{x} : \mid \mathbf{x} - \mathbf{y} \mid \leq \delta \wedge \mathbf{y} \in S \wedge \mathbf{x} \notin B_{\bullet} \setminus S\}$$

The delta coverage can be imagined to result from the application of an operator a_δ to B:

$$\overline{D}_{+\delta} = a_\delta(B).$$

The delta coverage $\overline{D}_{+\delta}$ is the "pod" of thickness δ covering the set B. The set obtained as union of B with its delta coverage is referred to a $B_{+\delta}$:

$$B_{+\delta} = B \cup a_\delta(B).$$

This operation can be reiterated, and the final set obtained after n iteration is referred to as $B_{+\delta}^n$:

$$B_{+\delta}^n = B_{+\delta}^{n-1} \cup a_\delta\left(B_{+\delta}^{n-1}\right),$$

where $B_{+\delta}^1 = B_{+\delta}$ and $B_{+\delta}^0 = B$.

The reiteration of a_δ is the base for the following definition:

Definition 8 (Conformal coverage) A conformal coverage $C_{+\delta}$ of thickness δ is the following limit[1]:

$$C_{+\delta} = \lim_{n \to +\infty} \bigcup_{i=0}^{n-1} a_\delta\left(B_{+\delta/n}^i\right).$$

In other words, the conformal coverage process can be thought of as obtained by applying infinitely many delta coverages, one after another, and each one of infinitesimal thickness. As shown in Fig. 13.1, in general, conformal coverage and delta coverage of the same thickness can lead to different bodies. Also, from the same figure it is clear that these transformations are not topological invariants.

Definition 9 (Conformal deposition) A conformal deposition of a film of material Z of thickness δ is the pair $(C_{+\delta}, Z)$.

Note that the material Z may be different from the materials X^i forming the original body.

The following theorem is simple:

Theorem 1 (Planarization) *Let d_S be the diameter of $\mathcal{B}$. Irrespective of the shape of $\mathcal{B}$, the conformal deposition of a layer for which $\delta \gg d_S$ produces a body whose*

---

[1] Such a limit is naively clear, but its rigorous specification should require elaboration many of technical details. The same considerations hold true anytime limits of this kind are introduced.

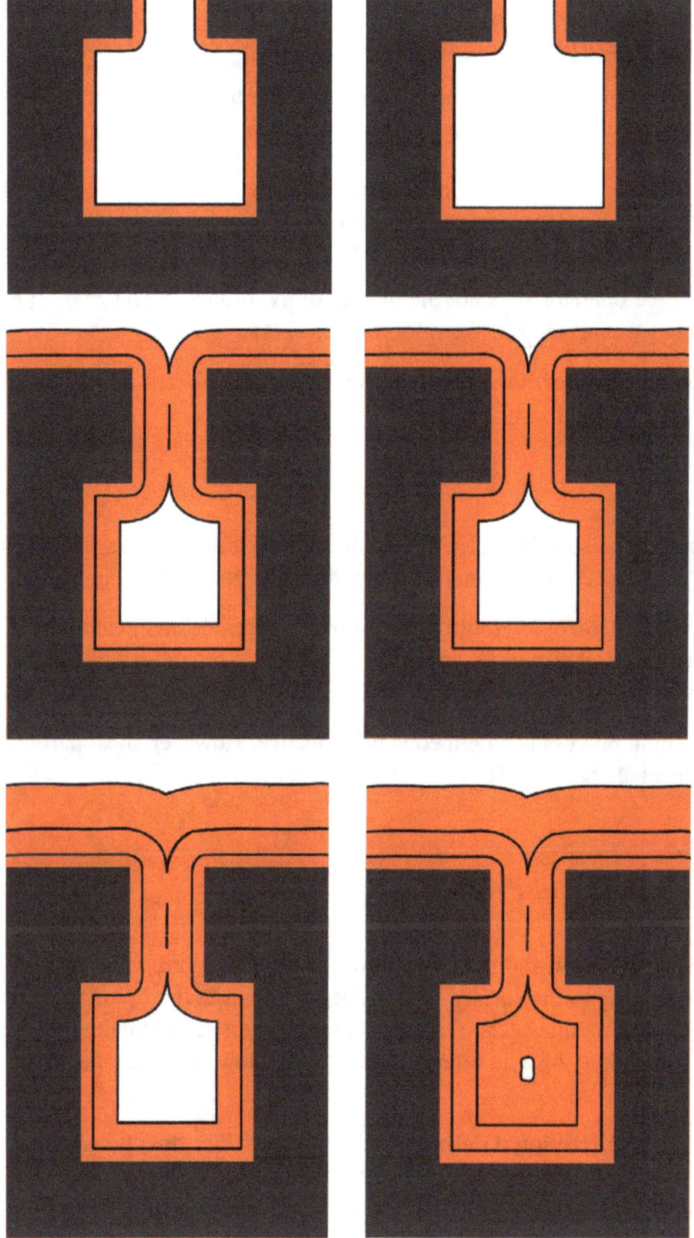

Fig. 13.1 The different behavior of conformal coverage (*left*) and delta coverage (*right*)

shape becomes progressively closer and loser to the spherical one as δ increases. If B is indefinitely extended in two directions, it undergoes a progressive planarization.

Isotropic etching can be defined in a way similar to that used for conformal coverage by introducing an operator s_δ, characterized by the following definitions:

$$s_\delta(B) = a_\delta(\mathbb{E}^3 \setminus B_\bullet)$$
$$B^n_{-\delta} = B^{n-1}_{-\delta} \setminus s_\delta(B^{n-1}_{-\delta})$$

In general, the operator s_δ is not the inverse of a_δ. In fact, $s_\delta(B) = a_{-\delta}(B)$ only if B is a simply connected set. We refer to s_δ as *delta depletion*.

Definition 10 (Conformal depletion) A conformal depletion $E_{-\delta}$ of thickness δ is the following limit:

$$E_{-\delta} = \lim_{n \to +\infty} \bigcup_{i=0}^{n-1} s_\delta\left(B^i_{-\delta/n}\right)$$

It is true that also in this case the conformal depletion and delta depletion s_δ of the same thickness can lead, in general, to different bodies (see Fig. 13.2).

Definition 11 (Isotropic etching) An isotropic (nonselective) etching is a conformal depletion of thickness δ applied to a body B, regardless of its constituting materials.

The etching process just defined is nonselective. However, most parts of etching are selective (see Sect. 13.3).

13.2.2 Directional Processes

Definition 12 (Shadowed surface along d) Let $\mathbf{d}$ be a unit vector in $\mathbb{E}^3$. For any point s belonging to the surface S, consider the straight line l_s starting from s and oriented as $\mathbf{d}$. If l_s intersects S, then s is said to be shadowed along $\mathbf{d}$. The set of all shadowed points of S is referred to as the shadowed surface $S_{\mathbf{d}}$ along $\mathbf{d}$ of the body B.

Definition 13 (Directional delta coverage along d) The directional delta coverage along $\mathbf{d}$ is the set of points lying on the straight lines directed along $\mathbf{d}$ and going from the exposed surface to the same surface shifted along $\mathbf{d}$ by an amount δ:

$$\overline{D}_{+\delta,+\mathbf{d}} = \{\mathbf{x} : \mathbf{x} = \mathbf{y} + a\mathbf{d} \wedge a \le \delta \wedge \mathbf{y} \in S \setminus S_{\mathbf{d}}\}$$

Figure 13.3 shows the results of applying to a given set, the delta coverage and the directional delta coverage along $\mathbf{d}$.

By using the directional delta coverage along δ instead of the delta coverage, it is possible to define the directional deposition:

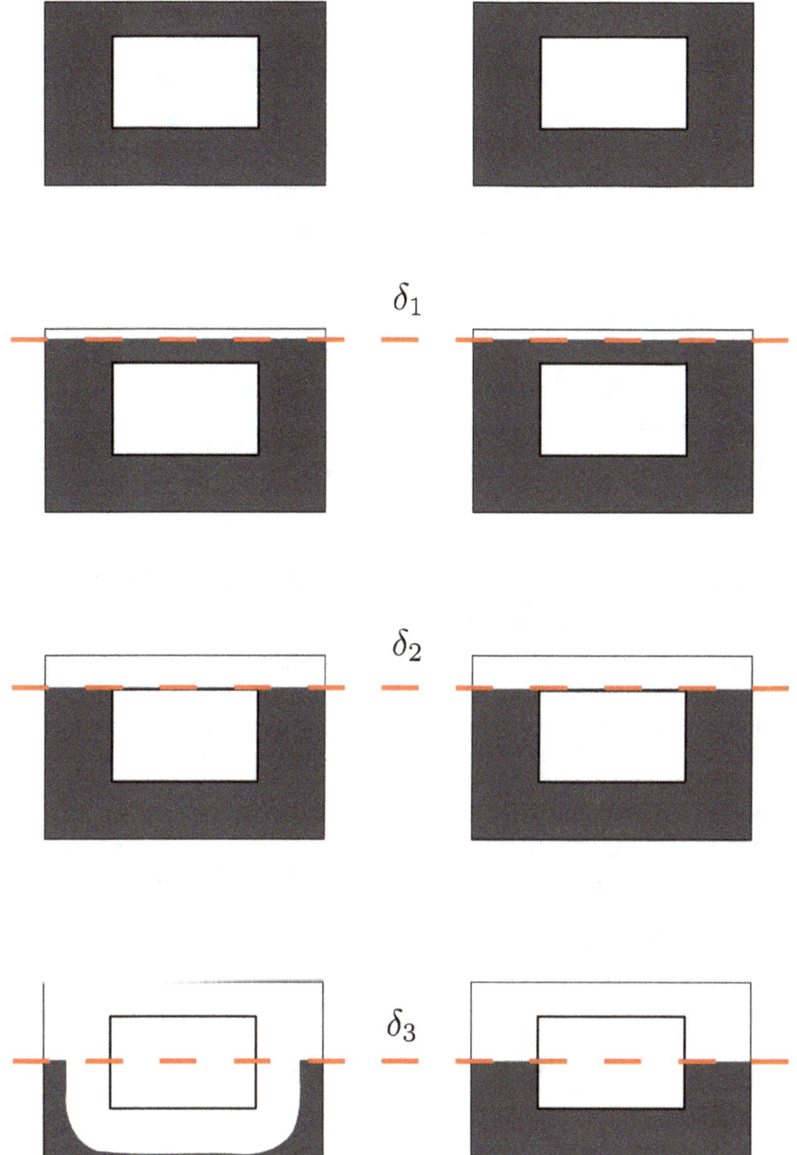

Fig. 13.2 The different behavior of conformal depletion (*left*) and delta depletion (*right*)

Definition 14 (Directional deposition) A directional deposition $C_{+\delta,+\mathbf{d}}$ of thickness δ along $\mathbf{d}$ is the following limit:

$$C_{+\delta,+\mathbf{d}} = \lim_{n \to +\infty} \bigcup_{i=0}^{n-1} a_{\delta,+\mathbf{d}}\big(B_{+\delta/n,+\mathbf{d}}^i\big),$$

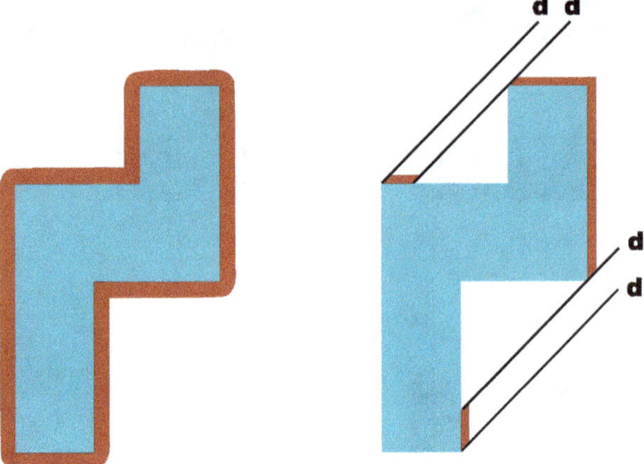

Fig. 13.3 Comparison between delta coverage (*left*) and directional delta coverage along **d** (*right*) of the same thickness δ

where $a_{\delta,+\mathbf{d}}$ is the operator associated with the directional delta coverage and $B^n_{+\delta,+\mathbf{d}} = B^{n-1}_{+\delta,+\mathbf{d}} \cup a_{\delta,+\mathbf{d}}(B^{n-1}_{+\delta,+\mathbf{d}})$.

Figure 13.4 shows an example of directional deposition. This process (as shown in the figure) can create holes inside the bodies.

Analogously, to define the directional depletion we introduce the directional delta depletion:

Definition 15 (Directional delta depletion along d)

$$\overline{D}_{+\delta,-\mathbf{d}} = \{\mathbf{x}: \ \mathbf{x} = \mathbf{y} - a\mathbf{d} \wedge a \le \delta \wedge \mathbf{y} \in S \setminus S_{\mathbf{d}}\}$$

Definition 16 (Directional etching along d) A directional etching $E_{-\delta,-\mathbf{d}}$ of thickness δ is the following limit:

$$E_{-\delta,-\mathbf{d}} = \lim_{n \to +\infty} \bigcup_{i=0}^{n-1} s_{\delta,-\mathbf{d}}\left(B^i_{-\delta/n,-\mathbf{d}}\right),$$

where $s_{\delta,-\mathbf{d}}$ is the operator associated with the directional delta depletion and $B^n_{-\delta,-\mathbf{d}} = B^{n-1}_{-\delta,-\mathbf{d}} \setminus s_{\delta,-\mathbf{d}}(B^{n-1}_{-\delta,-\mathbf{d}})$.

13.3 Processes Controlled by the Material

No indication has hitherto been given about the dependence of the same unit process on the material to which it is applied. In general, the effect depends on the material, and this difference is referred to in terms of selectivity.

Fig. 13.4 An example of directional deposition

Before introducing selectivity, we note that the surface S of a body $\mathcal{B}$ is, in general, composed of different materials. In particular, by using Proposition 1, it is possible to define $S_{\text{tot,X}}$, the subset of S_{tot} containing material X:

$$S_{\text{tot,X}} = \bigcup_{\text{X}^i = \text{X}} B_i^* \setminus \bigcup_{j,k} F_{j|k},$$

and then the (outer) surface S_{X} composed by material X:

$$S_{\text{X}} = S_{\text{tot,X}} \cap S.$$

Definition 17 (Selective delta depletion) For any body $\mathcal{B}$ with surface S, the selective delta depletion $s_{\delta,X}(B)$ of thickness δ is the set

$$s_{\delta,X}(B) = \{\mathbf{x} : \mid \mathbf{x} - \mathbf{y} \mid \leq \delta \wedge \mathbf{y} \in S_X \wedge \mathbf{x} \notin B_\bullet \setminus S_X\}.$$

In other words, the selective delta depletion is a delta depletion that affects S_X only, not the whole surface S. In general, $s_{\delta,X}(B)$ is not a connected set.

Definition 18 (Selectivity) Any process is said to be selective with respect to materials X when, applied to a composite body $\mathcal{B}$ with frontier S, it affects S_X only.

Definition 19 (Selective etching) A selective etching $E_{-\delta,X}$ of thickness δ is the following limit (in a sense to be defined):

$$E_{-\delta,X} = \lim_{n \to +\infty} \bigcup_{i=0}^{n-1} s_{\delta,X}\big(B^i_{-\delta/n,X}\big),$$

where

$$B_{-\delta/n,X} = \big(B \setminus s_{\delta,X}(B)\big) \cup \left(\bigcup_{X^j \neq X} B_j\right),$$

and the iterative process is defined as in the previous cases.

An example of selective etching is given in Fig. 13.5.

13.4 Abstract Technology in Concrete

Together with the above operations, one could certainly define other, more complicated (or perhaps, even more interesting) operations. What is of uppermost importance here is that *there already exist combinations of materials and processes of the silicon technology mimicking the ideal behaviors described above.*

The materials considered in this chapter are well known in silicon technology: A practical model for wafer is any body extending in two directions excessively more than the change of thickness resulting from the considered processes and such that the processes are actually carried out on only one of its major surfaces. Before undergoing any operations, such wafers have generally homogeneous chemical composition and high flatness and are referred to as substrates. In the typical situations considered herein, the substrates are slices of single-crystalline silicon. The other materials are poly-Si, SiO_2, Si_3N_4, and various metals (Al, Ti, Pt, Au, etc.).

The typical processes involved in the silicon technology are photolithography (for the definition of geometries), wet or gas-phase etchings, CVD, PVD planarization, doping (typically via ion implantation), oxidation, and diffusion. In a little more detail, the processes of interest here are as follows:

Fig. 13.5 An example of selective etching ("lift-off")

Lithography   requires a preliminary planarization of the surface with a resist, its patterning (i.e., the definition of a geometry) via the exposure through a mask to light, the selective etching of the exposed (or unexposed) resist, the selective etching first of the region not protected by the patterned resist, and eventually of this material. Resists are photoactive materials undergoing polymerization (or depolymerization) under illumination; the depolymerized region can thus be etched by a solvent to which the polymer is resistant (hence the name, *resist*), thus leaving a protective pattern on the surface.

Wet etchings   are usually isotropic and are used for their selectivity: HF_{aq} etches isotropically SiO_2 leaving unchanged silicon, and Si_3N_4; H_3PO_4 etches isotropically Si_3N_4 leaving unchanged silicon and SiO_2; $HF_{aq} + HNO_{3\,aq}$ etches Si leaving unchanged Si_3N_4 (but has poor selectivity with respect to SiO_2).

Table 13.1 Shape resulting after etching or growth processes

Process	Shape Conformal	Directional
Attack	Wet etching	Sputter etching
	⟵ Plasma etching	
		Reactive ion etching ⟶
	Oxidation (for Si)	
Growth	CVD	PVD
	Oxidation (of Si)	
Doping	Diffusion	Ion implantation

Sputter etching   is produced by momentum transfer from a beam to a target and results typically in nonselective directional etching. Selectivity can be imparted exploiting reactive ion etching.

Plasma etching   is a process whose degree of directionality can be tuned to the situation: via a suitable choice of the atmosphere, it can be used for the isotropic selective etching of Si, SiO_2 or Si_3N_4; it becomes progressively more directional and less selective applying a bias to the body ("target") with respect to the plasma.

Chemical vapor deposition   is the typical way for the conformal deposition that occurs when the growth is controlled by reactions occurring at the growing surface. Poly-Si grows well on SiO_2 but requires an SiO_2 buffer layer for the growth on Si_3N_4; conversely, Si_3N_4 is easily deposited on SiO_2. Silicon can be conformally covered by extremely uniform layers of SiO_2 with controlled thickness on the subnanometer range via *thermal oxidation*.

Physical vapor deposition   is the typical way for the deposition of metal films. The corresponding growth mode is reminiscent of, but not exactly the same as, the positive counterpart of directional etching (see Fig. 13.5).[2]

Oxidation   of silicon plays a special role; indeed, it results in the simultaneous conformal deposition of the surface with a layer of thickness t_{SiO_2} of SiO_2 and etching of the original material by an amount of thickness t_{Si}, with $t_{Si} \simeq t_{SiO_2}$. Oxidation can occur selectively protecting the silicon with pyrolitic S_3N_4.

Table 13.1 summarizes the shape (conformal or directional) resulting from the above processes.

---

[2] The conformal deposition of a metal layer (a fundamental step for the preparation of ICs with metal interconnects arranged in several layers) can only be done via chemical vapor deposition (CVD) from volatile precursors, like metal carbonyls or metallorganic monomers.

References

1. R.P. Feynman, Eng. Sci. **23**, 22 (1985)
2. H.J. Queisser, private communication (January 15, 2007)
3. K.E. Drexler, *Nanosystems – Molecular Machines, Manufacturing and Computation* (Wiley, New York, 1985)
4. K.E. Drexler, *Engines of Creation. The Coming Era of Nanotechnology* (Anchor Books, New York, 1986)
5. H. Rohrer (interviewed by P.S. Weiss), ACS Nano, **1**, 3 (2007)
6. A.M. Stoneham, Mater. Sci. Eng. C **23**, 235 (2003)
7. Anon., Nature **432**, 8 (2004)
8. A. Aviram, M. Ratner, Chem. Phys. Lett. **29**, 277 (1974)
9. M.A. Reed, MRS Bull. **26**(2), 113 (2001)
10. J.M. Tour, in *Stimulating Concepts in Chemistry*, ed. by F. Vögtle, J.F. Stoddart, M. Shibasaki (Wiley-VCH, Weinheim, 2000), p. 237
11. R.W. Keyes, in *Molecular Electronics and Molecular Electronic Devices*, ed. by K. Sienicki (CRC Press, Boca Raton, 1993), Chap. 1
12. J. Chen, M.A. Reed, A.M. Rawlett, J.M. Tour, Science **286**, 1550 (1999)
13. D.I. Gittins, D. Bethell, D.J. Schiffrin, R.J. Nichols, Nature **408**, 67 (2000)
14. J.E. Lilienfeld, US Patent 1745175, Application filed 8 October 1926, granted 18 January 1930
15. O. Heil, British Patent 439457, Application filed 4 March 1935, granted 6 December 1935
16. W. Shockley, IEEE Trans. Electron. Dev. **ED-23**, 597 (1976)
17. D. Khang, M.M. Atalla, IRE Solid-State Device Research Conference, Pittsburgh, 1960
18. H. Queisser, *The Conquest of the Microchip* (Harvard University Press, Cambridge, MA, 1988)
19. M.S. Malone, *The Microprocessor. A Biography* (Springer, New York, NY, 1995)
20. G.E. Moore, Phys. Today **44**(1), 83 (1991)
21. G.E. Moore, IEDM Technol. Dig. 11 (1975)
22. G.E. Moore, Electronics **38**, 114 (1965)
23. F. Faggin, T. Klein, L. Vadasz, IEEE Trans. Electron. Dev. **16**(2), 236 (1969)
24. F. Faggin, T. Klein, Solid State Electron. **13**, 1125 (1970)
25. L.L. Vadasz, A.S. Grove, T.A. Rowe, G.E. Moore, IEEE Spectrum **6**, 28 (1969)
26. J.A. Appels, M.M. Paffen, Philips Res. Rep. **6**, 157 (1971)
27. E. Kooi, J.G. van Lierop, W.H.C.G. Verkuijlen, R. de Werdt, Philips Res. Rep. **26**, 166 (1971)
28. F. Morandi, IEEE International Electron Device Meeting, 1969, p. 126
29. W.R. Hunter, T.C. Holloway, P.K. Chatterjee, A.F. Tasch Jr., IEEE Electron. Device Lett. **EDL 2**, 4 (1981)
30. A.D. Wilson, in *The Physics of VLSI*, ed. by J.C. Knights (American Institute of Physics, New York, 1984), p. 69
31. J. Birnbaum, R.S. Williams, Phys. Today **53**(1), 38 (2000)
32. E. Ross, *Future Fab Int.*, Issue 17 (2001), available at http://www.future-fab.com/

33. P. Gargini, *Semicon West* (July 2004), available at http://www.intel.com/research/silicon/nanotechnology.htm
34. Semiconductor Industry Association (SIA), International Technology Roadmap for Semiconductors. 2005 Edition, available at http://public.itrs.net
35. C. Rowen, Computer **35**(12), 29 (2002)
36. J. Henkel, Computer **36**(9), 119 (2003)
37. J.A. Hutchby, G.I. Bourianoff, V.V. Zhirnov, J.E. Brewer, IEEE Circuits Dev. Mag. **18**(2), 28 (2002)
38. G.F. Cerofolini, Appl. Phys. A **86**, 23 (2007)
39. A. Chiabrera, E. Di Zitti, F. Costa, G.M. Bisio, J. Phys. D: Appl. Phys. **22**, 1571 (1989)
40. R.W. Landauer, IBM J. Res. Dev. **5**, 183 (1961)
41. C.H. Bennett, IBM J. Res. Dev. **17**, 525 (1973)
42. R.W. Keyes, Science **195**, 1230 (1977)
43. R.W. Keyes, Proc. IEEE **69**, 267 (1981)
44. E. Fredkin, T. Toffoli, Int. J. Theor. Phys. **21**, 219 (1982)
45. C.H. Bennett, Int. J. Theor. Phys. **21**, 905 (1982)
46. R. Landauer, Nature **335**, 779 (1988)
47. W.H. Zurek, Nature **341**, 119 (1989)
48. R.P. Feynman, in *Lectures on Computation*, ed. by J.G. Hey, R.W. Allen (Addison Wesley, Reading, MA, 1995), Chap. 5, p. 137
49. S. Lloyd, Nature **406**, 1047 (2000)
50. J.D. Meindl, J.A. Davis, IEEE J. Solid State Circuits **35**, 1515 (2000)
51. J.A. Davis, R. Venkatesan, A. Kaloyeros, M. Beylansky, S.J. Souri, K. Banerjee, K.C. Saraswat, A. Rahman, R. Reif, J.D. Meindl, Proc. IEEE **89**, 305 (2001)
52. V.V. Zhirnov, R.K. Cavin, J.A. Hutchby, G.I. Bourianoff, Proc. IEEE **91**, 1934 (2003)
53. P.A. Packan, Science **285**, 2079 (1999)
54. P.S. Peercy, Nature **406**, 1023 (2000)
55. R.W. Keyes, Proc. IEEE **89**, 227 (2001)
56. D.J. Frank, R.H. Dennard, E. Nowak, P.M. Solomon, Y. Taur, H.-S.P. Wong, Proc. IEEE **89**, 259 (2001)
57. M. Stoneham, Mater. Today**11**(9), 32 (2008)
58. J.A. Wheeler, W.H. Zurek (Eds.), *Quantum Theory and Measurement* (Princeton University Press, Princeton, NJ, 1983)
59. Y. Aharonov, D. Bohm, Phys. Rev. **122** 1649 (1961)
60. C.A. Stafford, D.M. Cardamone, S. Mazumdar, Nanotechnology **18**, 424014 (2007)
61. R.F. O'Connell, IEEE Trans. Nanotechnol. **4**, 77 (2005)
62. D. Cohen, B. Horovitz, J. Phys. A: Math. Theor. **40**, 12281 (2007)
63. T. Toffoli, in *Encyclopedia of Electrical and Electronics Engineering*, ed. by J.G. Webster (Wiley, New York, NY, 1988), p. 455
64. J.-Y. Girard, in *Advances in Linear Logic*, ed. by J.-Y. Girard, Y. Lafont, L. Regnier (Cambridge University Press, 1993), pp. 1–42
65. C.S. Lent, Science **288**, 1597 (2000)
66. S.G. Brush, *The Kind of Motion We call Heat. Book 2: Statistical Physics and Irreversible Processes* (North Holland, Amsterdam, 1976), pp. 589–590
67. L. Szilard, Z. f. Phys. **53**, 840 (1929); English translation reprinted in [58, pp. 539–548]
68. L. Brillouin, J. Appl. Phys. **22**, 334 (1951)
69. L.G.M. Gordon, Found. Phys. **11**, 103 (1981)
70. L.G.M. Gordon, Found. Phys. **13**, 989 (1983)
71. C.M. Evans, Phys. Rev. Lett. **64**, 2111 (1990)
72. E.P. Gyftopoulos, Physica A **307**, 405, 421 (2002)
73. J. Bub, arXiv:quant-ph/0203017v1 (5 March 2002)
74. C.H. Bennett, Stud. Hist. Philos. Mod. Phys. **34**, 501 (2003)
75. M. Devereux, Entropy **6**, 102 (2004)
76. C. Van den Broeck, P. Meurs, R. Kawai, New J. Phys. **7**(10), 1 (2005)
77. M. Toda, R. Kubo, N. Saitô, *Statistical Physics I* (Springer, Berlin, 1983), Sect. 2.7

78. G.F. Cerofolini, L. Meda, Phys. Rev. B **36**, 5131 (1987)
79. G.F. Cerofolini, L. Meda, C. Volpones, J. Appl. Phys. **63**, 4911 (1988)
80. G.F. Cerofolini, L. Meda, C. Volpones, Mater. Res. Soc. Symp. Proc. **128**, 181 (1989)
81. M. Werner, R. Locher, Rep. Prog. Phys. **61**, 1665 (1998)
82. G.F. Cerofolini, Appl. Phys. A **51**, 467 (1990)
83. D. Twerenbold, Rep. Prog. Phys. **59**, 349 (1996)
84. N.E. Booth, B. Cabrera, E. Fiorini, Annu. Rev. Nucl. Part. Sci. **46**, 471 (1996)
85. F. Gatti, Nucl. Phys. B: Proc. Suppl. **78**, 99, (1999)
86. R.L. McCreery, Chem. Mater. **16**, 4477 (2004)
87. R.W. Keyes, private communication (July 30, 2001)
88. R.W. Keyes, J. Phys.: Condens. Matter **18**, S703 (2006)
89. G. Charpak, Annu. Rev. Nucl. Sci. **20**, 195 (1970)
90. G. Charpak, F. Sauli, Nucl. Instrum. Methods **162**, 405 (1979)
91. C.W. Fabjan, H.G. Fischer, Rep. Prog. Phys. **43**, 1003 (1980)
92. G.F. Cerofolini, G. Ferla, A. Foglio Para, Nucl. Instrum. Methods **169**, 125 (1980)
93. J.R. Heath, P.J. Kuekes, G.S. Snider, R.S. Williams, Science **280**, 1716 (1998)
94. M.M. Ziegler, M.R. Stan, IEEE Trans. Nanotechnol. **2**, 217 (2003)
95. A. DeHon, IEEE Trans. Nanotechnol. **2**, 23 (2003)
96. P.M. Mendes, A.H. Flood, J.F. Stoddart, Appl. Phys. A **80**, 1197 (2005)
97. P.J. Kuekes, D.R. Stewart, R.S. Williams, J. Appl. Phys. **97**, 034301 (2005)
98. J.C. Love, L.A. Estroff, J.K. Kriebel, R.G. Nu, G.M. Whitesides, Chem. Rev. **105**, 1103 (2005)
99. M.A. Reed, J. Chen, A.M. Rawlett, D.W. Price, J.M. Tour, Appl. Phys. Lett. **78**, 3735 (2001)
100. Y. Luo, C.P. Collier, J.O. Jeppesen, K.A. Nielsen, E. Delonno, G. Ho, J. Perkins, H.-R. Tseng, T. Yamamoto, J.F. Stoddart, J.R. Heath, Chem. Phys. Chem. **3**, 519 (2002)
101. R.F. Service, Science **302**, 556 (2003)
102. D.R. Stewart, D.A.A. Ohlberg, P. Beck, Y. Chen, R.S. Williams, J.O. Jeppesen, K.A. Nielsen, J.F. Stoddart, Nano Lett. **4**, 133 (2004)
103. C.N. Lau, D.R. Stewart, R.S. Williams, D. Bockrath, Nano Lett. **4**, 569 (2004)
104. N.B. Zhitenev, W. Jiang, A. Erbe, Z. Bao, E. Garfunkel, D.M. Tennant, R.A. Cirelli, Nanotechnology **17**, 1272 (2006)
105. M.P. Stewart, F. Maya, D.V. Kosynkin, S.M. Dirk, J.J. Stapleton, C.L. McGuiness, D.L. Allara, J.M. Tour, J. Am. Chem. Soc. **126**, 370 (2004)
106. G.F. Cerofolini, G. Ferla, J. Nanoparticle Res. **4**, 185 (2002)
107. J.E. Green, J.W. Choi, A. Boukai, Y. Bunimovich, E. Johnston–Halperin, E. Delonno, Y. Luo, B.A. Sheriff, K. Xu, Y.S. Shin, H.-R. Tseng, J.F. Stoddart, J.R. Heath, Nature **445**, 414 (2007)
108. H.B. Akkerman, P.W.M. Blom, D.M. de Leeuw, B. de Boer, Nature **441**, 69 (2006)
109. G.F. Cerofolini, G. Arena, M. Camalleri, C. Galati, S. Reina, L. Renna, D. Mascolo, V. Nosik, Microelectr. Eng. **81**, 405 (2005)
110. G.F. Cerofolini, G. Arena, M. Camalleri, C. Galati, S. Reina, L. Renna, D. Mascolo, Nanotechnology **16**, 1040 (2005)
111. G.F. Cerofolini, Nanotechnol. E-Newslett. **7**(3) 5 (2005)
112. G.F. Cerofolini, L. Meda, *Physical Chemistry of, in and on Silicon* (Springer, Berlin, 1989), Chap. 10
113. G.F. Cerofolini, D. Mascolo, in *Nanotechnology for Electronic Materials and Devices*, ed. by E. Gusev, A. Korkin, J. Labanowski, S. Luryi (Springer, New York, 2006), Chap. 1, p. 1
114. M.H.R. Lankhorst, B.W.S.M.M. Ketelaars, R.A.M. Wolters, Nat. Mater. **4**, 347 (2005)
115. D.B. Strukov, G.S. Snider, D.R. Stewart, R.S. Williams, Nature **453**, 80 (2008)
116. Y.-K. Choi, T.-J. King, C. Hu, IEEE Trans. Electron. Devices **49**, 436 (2002)
117. G.A. Garfunkel, M.B. Weissman, J. Vac. Sci. Technol. B **8**, 1087 (1990)
118. D.C. Flanders, A.E. White, J. Vac. Sci. Technol. **19**, 892 (1981)
119. D.B. Gates, Q.B. Xu, M. Stewart, D. Ryan, C.G. Willson, G.M. Whitesides, Chem. Rev. **105**, 1171 (2005)
120. D. Natelson, R.L. Willett, K.W. West, L.N. Pfeiffer, Appl. Phys. Lett. **77**, 1991 (2000)

121. N.A. Melosh, A. Boukai, F. Diana, B. Gerardot, A. Badolato, J.R. Heath, Science **300**, 112 (2003)
122. D. Wang, B.A. Sheriff, M. McAlpine, J.R. Heath, Nano Res. **1**, 9 (2008)
123. Y.-K. Choi, J. Zhu, J. Grunes, J. Bokor, G.A. Somorjai, J. Phys. Chem. B **107**, 3340 (2003)
124. Y.-K. Choi, J.S. Lee, J. Zhu, G.A. Somorjai, L.P. Lee, J. Bokor, J. Vac. Sci. Technol. **21**, 2951 (2003)
125. D.C. Flanders, N.N. Efremow, J. Vac. Sci. Technol. B **1**, 1105 (1983)
126. G.F. Cerofolini, D. Mascolo, Semicond. Sci. Technol. **21**, 1315 (2006)
127. G.F. Cerofolini, Appl. Phys. A **86**, 31 (2007)
128. L.T. Canham, Appl. Phys. Lett. **57**, 1046 (1990)
129. M. Roukes, Sci. Am. Rep. **17**(3), 4 (2007)
130. M. Forshaw, R. Stadler, D. Crawley, K. Nikolic, Nanotechnology **15**, S220 (2004)
131. Y. Huang, X Duan, Y. Cui, L.J. Lauhon, K.-H. Kim, C.M. Lieber, Science **294**, 1313 (2001)
132. Z. Zhong, D. Wang, Y. Cui, M.W. Bockrath, C.M. Lieber, Science **302**, 1377 (2003)
133. A. DeHon, P. Lincoln, J.E. Savage, IEEE Trans. Nanotechnol. **2**, 165 (2003)
134. R. Alley, M. Cumbie, R. Enck, D. Huang, P. Kornilovitch, S. Ramamoorthi, J. Wu, X. Yang, MST News No. 4/06, 8 (2006)
135. K.K. Likharev, in *Nano and Giga Challenges in Microelectronics*, ed. by J. Greer, A. Korkin, J. Labanowsky (Elsevier, Amsterdam, 2003), p. 27
136. K. Likharev, D.B. Strukov, in *Introducing Molecular Electronics*, ed. by G. Cuniberti, G. Fagas, K. Richter (Springer, Berlin, 2005), Chap. 16, p. 447
137. D.B. Strukov, K. Likharev, Nanotechnology **16**, 137 (2005)
138. R. Beckman, E. Johnston-Halperin, Y. Luo, J.E. Green, J.R. Heath, Science **310**, 465 (2005)
139. R. Amerson, R.J. Carter, W.B. Culbertson, P. Kuekes, G. Snider, Proceedings of the IEEE Symposium on FPGAs for Custom Computing Machines, 1995, p. 32
140. W. Wu, G.-Y. Jung, D.L. Olynick, J. Strasnicky, Z. Li, X. Li, D.A.A. Ohlberg, Y. Chen, S.-Y. Wang, J.A. Liddle, W.M. Tong, R.S. Williams, Appl. Phys. A **80**, 1173 (2005)
141. D.K. Aswal, S. Lenfant, D. Guerin, J.V. Yakhmi, D. Vuillaume, Analytica Chimica Acta **568**, 84 (2006)
142. S. Ossicini, L. Pavesi, F. Priolo, *Light Emitting Silicon for Microphotonics* (Springer, Berlin, 2003)
143. F. Koch, Mater. Res. Soc. Symp. Proc. **298**, 319 (1993)
144. T. Shimizu-Iwayama, S. Nakao, K. Saitoh, Appl. Phys. Lett. **65**, 1814 (1994)
145. P. Mutti, G. Ghislotti, S. Bertoni, L. Bonoldi, G.F. Cerofolini, L. Meda, E. Grilli, M. Guzzi, Appl. Phys. Lett. **66**, 851 (1995)
146. G.F. Cerofolini, L. Meda, D. Bisero, F. Corni, G. Ottaviani, R. Tonini, Mater. Sci. Eng. B **36**, 108 (1996)
147. R.Rölver, S. Brüninghoff, M. Först, B. Spangenberg, H. Kurz, J. Vac. Sci. Technol. B **23**, 3214 (2005)
148. S. Nihonyanagi, Y. Kanemitsua, Appl. Phys. Lett. **85**, 5721 (2004)
149. N. Pauc, V. Calvo, J. Eymery, F. Fournel, N. Magnea, Phys. Rev. B **72**, 205324 (2005)
150. N. Pauc, V. Calvo, J. Eymery, F. Fournel, N. Magnea, Phys. Rev. B **72**, 205325 (2005)
151. S. Saito, D. Hisamoto, H. Shimizu,H. Hamamura, R. Tsuchiya, Y. Matsui, T. Mine, T. Arai, N. Sugii, K. Torii, S. Kimura, T. Onai, Jap. J. Appl. Phys. **45**, L679 (2006)
152. Y.-K. Hong, N.W. Song, J.H. Bahng, S. Lee, J.-Y. Kooz, T.S. Park, M. Yoon, J.-Y. Yi, J. Kor. Phys. Soc. **48**, 974 (2006)
153. L.E. Bell, Science **321**, 1457 (2008)
154. A.I. Hochbaum, R. Chen, R.D. Delgado, W. Liang, E.C. Garnett, M. Najarian, A. Majumdar, P. Yang, Nature **451**, 163 (2008)
155. A.I. Boukai, Y. Bunimovich, J. Tahir-Kheli, J.-K. Yu, W.A. Goddard, J.R. Heath, Nature **451**, 168 (2008)
156. J.C. Love, L.A. Estroff, J.K. Kriebel , R.G. Nuzzo, G.M. Whitesides, Chem. Rev. **105**, 1103 (2005)
157. H. Yu, Y. Luo, K. Beverly, J.F. Stoddart, H.-R. Tseng, J.R. Heath, Angew. Chem. Int. Ed. **42**, 5706 (2003)

158. D.M. Adams, L. Brus, C.E.D. Chidsey, S. Creager, C. Creutz, C.R. Kagan, P.V. Kamat, M. Lieberman, S. Lindsay, R.A. Marcus, R.M. Metzger, M.E. Michel-Beyerle, J.R. Miller, M.D. Newton, D.R. Rolison, O. Sankey, K.S. Schanze, J. Yardley, X. Zhu, J. Phys. Chem. B **107**, 6668 (2003)
159. L. Baldi, G.F. Cerofolini, G. Ferla, G. Frigerio, Phys. Stat. Sol. (a) **48**, 523 (1978)
160. L. Baldi, G.F. Cerofolini, G. Ferla, J. Electrochem. Soc. **127**, 125 (1980)
161. G.F. Cerofolini, M.L. Polignano, J. Appl. Phys. **55**, 579 (1984)
162. M.L. Polignano, G.F. Cerofolini, H. Bender, C. Claeys, J. Appl. Phys. **64**, 869 (1988)
163. P. Cappelletti, G.F. Cerofolini, M.L. Polignano, J. Appl. Phys. **57**, 646 (1985)
164. G.F. Cerofolini, M.L. Polignano, J. Appl. Phys. **55**, 3823 (1984)
165. G.F. Cerofolini, M.L. Polignano, Appl. Phys. A **50**, 273 (1990)
166. G.F. Cerofolini, Phys. Stat. Sol. (a) **102**, 345 (1987)
167. R. Landauer, IBM J. Res. Dev. **1**, 223 (1957)
168. R.Büttiker, Y. Imry, R. Landauer, S. Pinhas, Phys. Rev. B **31**, 6207 (1985)
169. F. Schwable, *Quantum Mechanics*, 2nd edn. (Springer, Berlin, 1998)
170. S. Flügge, *Practical Quantum Mechanics* (Springer, Berlin, 1994)
171. A. Salomon, T. Boecking, C.K. Chan, F. Amy, O. Girshevitz, D. Cahen, A. Kahn, Phys. Rev. Lett. **95**, 266807 (2005)
172. O. Seitz, T. Böcking, A. Salomon, J.J. Gooding, D. Cahen, Langmuir **22**, 6915 (2006)
173. G.F. Cerofolini, A. Giussani, F. Carone Fabiani, A. Modelli, D. Mascolo, D. Ruggiero, D. Narducci, E. Romano, Surf. Interface Anal. **39**, 836 (2007)
174. W. Wang, T. Lee, M.A. Reed, J. Phys. Chem. B **108**, 18398 (2004)
175. W. Wang, T. Lee, I. Kretzschmar, M.A. Reed, Nano Lett. **4**, 643 (2004)
176. G. Fagas, J.C. Greer, Nanotechnology **18**, 424010 (2007)
177. J. He, B. Chen, A.K. Flatt, J.J. Stephenson, C.D. Doyle, J.M. Tour, Nat. Mater. **5**, 63 (2006)
178. J.M. Seminario, A.G. Zacarias, J.M. Tour, J. Am. Chem. Soc. **122**, 3015 (2000)
179. Y. Ikenoue, N. Uotani, A.O. Patil, F. Wudl, A.J. Heeger, Synth. Met. **30**, 305 (1989)
180. E.H. Hauge, J.A. Støvneng, Rev. Mod. Phys. **61**, 917 (1989)
181. P. Pfeifer, J. Frölich, Rev. Mod. Phys. **67**, 761 (1995)
182. M.A. Brook, *Silicon in Organic, Organometallic and Polymer Chemistry* (Wiley, Chichester, 2000)
183. T.J. Barton, P. Boudjouk, in *Silicon-Based Polymer Science. A Comprehensive Resource*, ed. by J.M. Zeigler, F.W. Gordon Fearon (American Chemical Society, Washington, DC, 1990), Chap. 1, p. 3
184. R. Walsh, Acc. Chem. Res. **14**, 246 (1981)
185. R. West, M.J. Fink, J. Michl, Science **214**, 1343 (1981)
186. T. Tsumuraya, S.A. Batcheller, S. Masamune, Angew. Chem. Int. Ed. **30**, 902 (1991)
187. E.J. Buehler, J.J. Boland, Science **290**, 506 (2000)
188. H.J.W. Zandvliet, Rev. Mod. Phys. **72**, 593 (2000)
189. S. Ciraci, H. Wagner, Phys. Rev. B **27**, 5180 (1983)
190. Y.J. Chabal, S.B. Christman, Phys. Rev. B **29**, 6974 (1984)
191. A. Bilic, J.R. Reimers, N.S. Hush, Surf. Rev. Lett. **11**, 185 (2004)
192. G.J. Pietsch, G.S. Higashi, Y.J. Chabal, Appl. Phys. Lett. **64**, 3115 (1994)
193. G.J. Pietsch, Y.J. Chabal, G.S. Higashi, J. Appl. Phys. **78**, 1650 (1995)
194. M. Morita, T. Ohmi, E. Hasegawa, M. Kawakami, M. Ohwada, J. Appl. Phys. **68**, 1272 (1990)
195. G.F. Cerofolini, G. La Bruna, L. Meda, Appl. Surf. Sci. **93**, 255 (1996)
196. F.J. Himpsel, F.R. McFeely, A. Taleb Ibrahimi, J.A. Yarnoff, G. Hollinger, Phys. Rev. B **38**, 6084 (1988)
197. P. Singer, Semicond. Int. **18**(10), 88 (1995)
198. M.C. Desjonquères, D. Spanjard, *Concepts in Surface Physics* (Springer, Berlin, 1996), Chap. 5, pp. 254–266
199. H.N. Waltenburg, J.T. Yates, Chem. Rev. **95**, 1589 (1995)
200. T. Umeda, S. Yamasaki, M. Nishizawa, T. Yasuda, K. Tanaka, Appl. Surf. Sci. **162–163**, 299 (2000)

201. Y.I. Chabal, K. Raghavachari, X. Zhang, E. Garfunkel, Phys. Rev. B **66**, 181315(R) (2002)
202. L.C.P.M. de Smet, H. Zuilhof, E.J.R. Sudhölter, L.H. Lie, A. Houlton, B.R. Horrocks, J. Phys. Chem. B **109**, 12020 (2005)
203. G.F. Cerofolini, C. Galati, S. Reina, L. Renna, N. Spinella, G.G. CondorelliPhys. Rev. B **74**, 235407 (2006)
204. T. Aoyama, K. Goto, T. Yamazaki, T. Ito, J. Vac. Sci. Technol. A **14**, 2909 (1996)
205. M. Terashi, J. Kuge, M. Shinohara, D. Shoji, M. Niwano, Appl. Surf. Sci. **130/132**, 260 (1998)
206. Y. Morita, H. Tokumoto, Appl. Phys. Lett. **67**, 2654 (1995)
207. G.S. Higashi, Y.J. Chabal, G.W. Trucks, K. Raghavachari, Appl. Phys. Lett. **56**, 656 (1990)
208. H. Ubara, T. Imura, A. Hiraki, Solid State Commun. **50**, 673 (1984)
209. G.W. Trucks, K. Raghavachari, G.S. Higashi, Y.J. Chabal, Phys. Rev. Lett. **65**, 504 (1990)
210. F.A. Cotton, G. Wilkinson, *Advanced Inorganic Chemistry*, 5th edn. (Wiley, New York, NY, 1988)
211. J. March, *Advanced Organic Chemistry*, 4th edn. (Wiley, New York, NY, 1992)
212. N. Miki, H. Kikuyama, I. Kawanabe, M. Miyashita, T. Ohmi, IEEE Trans. Electron. Dev. **37**, 107 (1990)
213. G.F. Cerofolini, L. Meda, Appl. Surf. Sci. **89**, 351 (1995)
214. G.F. Cerofolini, N. Re, in *Fundamental Aspects of Ultrathin Dielectrics on Si Based Devices*, ed. by E. Garfunkel, E. Gusev, A. Vul (Kluwer, Dordrecht, 1998), pp. 117–129
215. G.F. Cerofolini, Appl. Surf. Sci. **133**, 108 (1998)
216. Y. Kumagai, K. Namba, T. Komeda, Y. Nishioka, J. Vac. Sci. Technol. A **16**, 1775 (1998)
217. G.F. Cerofolini, C. Galati S. Reina, L. Renna, N. Spinella, D. Jones, V. Palermo, Phys. Rev. B **72**, 125431 (2005)
218. G.F. Cerofolini, J. Colloid Interface Sci. **167**, 453 (1994)
219. G.F. Cerofolini, L. Meda, R. Falster, *Semiconductor Silicon 1994*, ed. by H.R. Huff, W. Bergholz, K. Sumino (The Electrochemical Society, Pennington, NJ, 1994), p. 379
220. G.F. Cerofolini, in *Silicon for the Chemical Industry III*, ed. by H.A. Oye, H.M. Rong, B. Ceccaroli, L. Nygaard, J.K. Tuset (Tapir, Trondheim, 1996), p. 117
221. V. Lehmann, Mater. Lett. **28**, 245 (1996)
222. L.T. Canham, A.J. Groszek, J. Appl. Phys. **72**, 1558 (1992)
223. P. Gupta, A.C. Dillon, A.S. Bracker, S.M. George, Surf. Sci. **245**, 360 (1991)
224. G. Smestad, M. Kunst, C. Vial, Solar Energy Mater. Solar Cells **26**, 277 (1992)
225. B. Hamilton, Semicond. Sci. Technol. **10**, 1187 (1995)
226. C. Steinem, A. Janshoff, V.S.Y. Lin, N.H. Volcker, M.R. Ghadiri, Tetrahedron **60**, 11259 (2004)
227. L.T. Canham, Adv. Mater. **7**, 1033 (1995)
228. M. Bjorkqvist, J. Salonen, E. Laine, Appl. Surf. Sci. **222**, 269 (2004)
229. R. Voicu, R. Boukherroub, V. Bartzoka, T. Ward, J.T.C. Wojtyk, D.D.M. Wayner, Langmuir **20**, 11713 (2004)
230. G.F. Cerofolini, G. Calzolari, F. Corni, S. Frabboni, C. Nobili, G. Ottaviani, R. Tonini, Phys. Rev. B **61**, 10183 (2000)
231. G.F. Cerofolini, F. Corni, S. Frabboni, C. Nobili, G. Ottaviani, R. Tonini, Mater. Sci. Eng. R **27**, 1 (2000)
232. V. Raineri, M. Saggio, E. Rimini, J. Mater. Res. **15**, 1 (2000)
233. S. Frabboni, F. Corni, C. Nobili, R. Tonini, G. Ottaviani, Phys. Rev. B **69**, 165209 (2004)
234. D.J. Eaglesham, A.E. White, L.C. Feldman, N. Moriya, D.C. Jacobson, Phys. Rev. Lett. **70**, 1643 (1993)
235. D.M. Follstaedt, Appl. Phys. Lett. **62**, 1116 (1993)
236. A.A. Stekolnikov, F. Bechstedt, Phys. Rev. B **72**, 125326 (2005)
237. T. Aoyama, K. Goto, T. Yamazaki, T. Ito, J. Vac. Sci. Technol. A **14**, 2909 (1996)
238. G.F. Cerofolini, C. Galati S. Reina, L. Renna, N. Spinella, D. Jones, V. Palermo, Phys. Rev. B **72**, 125431 (2005)
239. S.J. Pearton, J.W. Corbett, M. Stavola, *Hydrogen in Crystalline Semiconductors* (Springer, Berlin, 1992)

240. E. Romano, D. Narducci, F. Corni, S. Frabboni, G. Ottaviani, R. Tonini, G.F. Cerofolini, Mater. Sci. Eng. B **159–160**, 173 (2009)
241. J. Narayan, O.W. Holland, W.H. Christie, J.J. Wortman, J. Appl. Phys. **57**, 2709 (1985)
242. S. Boninelli, G. Impellizzeri, S. Mirabella, F. Priolo, E. Napolitani, N. Cherkashin, F. Cristiano, Appl. Phys. Lett. **93**, 061906 (2008)
243. R.S. Hockett, Appl. Phys. Lett. **54**, 1793 (1989)
244. L.S. Adam, M.E. Law, K.S. Jones, O. Dokumaci, C.S. Murthy, S. Hegde, J. Appl. Phys. **87**, 2282 (2000)
245. G.F. Cerofolini, Heter. Chem. Rev. **1**, 183 (1994)
246. S.F. Bent, Surf. Sci. **500**, 879 (2002)
247. G. Cleland, B.R. Horrocks, A Houlton, J. Chem. Soc. Faraday Trans. **91**, 4001 (1995)
248. A. Ulman, Adv. Mater. **2**, 573 (1990)
249. C.A. Roth, Ind. Eng. Chem. Prod. Res. Dev. **11**, 134 (1972)
250. J.M. Buriak, Chem. Rev. **102**, 1271 (2002)
251. F.A. Cotton, G. Wilkinson, *Advanced Inorganic Chemistry*, 5th edn. (Wiley, New York, NY, 1988)
252. J.S. Hovis, R.J. Hamers, J. Phys. Chem. B **101**, 9581 (1997)
253. H. Liu, R.J. Hamers, Surf. Sci. **416**, 354 (1998)
254. M.P. Schwartz, M.D. Ellison, S.K. Coulter, J.S. Hovis, R.J. Hamers, J. Am. Chem. Soc. **122**, 8529 (2000)
255. M.P. Schwartz, R.J. Hamers, Surf. Sci. **515**, 75 (2002)
256. J. Terry, M.R. Lindford, C. Wigren, R. Cao, P. Pianetta, C.E.D. Chidsey, J. Appl. Phys. **85**, 213 (1999)
257. A.B. Sieval, A.L. Demirel, J.W.M. Nissink, J.H. van der Maas, W.H. de Jeu, H. Zuilhof, E.J.R. Sudhölter, Langmuir **14**, 1759 (1998)
258. A.B. Sieval, V. Vleemimg, H. Zuilhof, E.J.R. Sudhölter, Langmuir **15**, 8288 (1999)
259. A. Scandurra, L. Renna, G. Cerofolini, S. Pignataro, Surf. Interface Anal. **34**, 777 (2002)
260. A. Lehner, G. Steinhoff, M.S. Brandt, M. Eickhoff, M. Stutzmann, J. Appl. Phys. **94**, 2289 (2003)
261. M. Kosuri, H. Gerung, Q. Li, S.M. Han, B.C. Bunker, T.M. Mayer, Langmuir **19**, 9315 (2003)
262. A.B. Sieval, R. Opitz, H.P.A. Maas, M.G. Schoeman, G. Meijer, F.J. Vergeldt, H. Zuilhof, E.J.R. Sudhölter, Langmuir **16**, 10359 (2000)
263. G.F. Cerofolini, C. Galati, S. Reina, L. Renna, Mater. Sci. Eng. C **23**, 253 (2003)
264. G.F. Cerofolini, C. Galati, S. Reina, L. Renna, Semicond. Sci. Technol. **18**, 423 (2003)
265. G.F. Cerofolini, C. Galati, S. Reina, L. Renna, O. Viscuso, G.G. Condorelli, I.L. Fragalà, Mater. Sci. Eng. C **23**, 989 (2003)
266. G.F. Cerofolini, C. Galati, S. Reina, L. Renna, Appl. Phys. A **80**, 161 (2004)
267. G.F. Cerofolini, C. Galati, S. Reina, L. Renna, Surf. Interface Anal. **36**, 71 (2004)
268. G.F. Cerofolini, C. Galati, S. Reina, L. Renna, G.G. Condorelli, I.L. Fragalà, G. Giorgi, A. Sgamellotti, N. Re, Appl. Surf. Sci. **246**, 52 (2005)
269. G.F. Cerofolini, C. Galati, S. Reina, L. Renna, Surf. Interface Anal. **38**, 126 (2006)
270. M. Woods, S. Carlsson, Q. Hong, S.N. Patole, L.H. Lie, A. Houlton, B.R. Horrocks, J. Phys. Chem. B **109**, 24035 (2005)
271. C. Coletti, A. Marrone, G. Giorgi, A. Sgamellotti, G.F. Cerofolini, N. Re, Langmuir **22**, 9949 (2006)
272. S.W. Howell, S.M. Dirk, K. Childs, H. Pang, M. Blain, R.J. Simonson, J.M. Tour, D.R. Wheeler, Nanotechnology **16**, 754 (2005)
273. G.F. Cerofolini, C. Galati, S. Reina, L. Renna, P. Ward, Appl. Phys. A **81**, 187 (2005)
274. G.F. Cerofolini, Mater. Sci. Semicond. Process. **5**, 265 (2003)
275. A. Hjelmfelt, J. Ross, Physica D **84**, 180 (1995)
276. S.J. Teichner, G.A. Nicolan, M.A. Vicarini, G.E.E. Gardes, Adv. Colloid Interface Sci. **5**, 245 (1976)
277. H.D. Gesser, P.C. Goswami, Chem. Rev. **89**, 765 (1989)
278. J. Fricke, A. Emmerling, Adv. Mater. **3**, 504 (1991)
279. S.S. Kistler, Nature **127**, 741 (1931)

280. L.L. Hench, J.K. West, Chem. Rev. **90**, 33 (1990)
281. S.S. Prakash, C.J. Brinker, A.J. Hurd, S.M. Rao, Nature **374**, 439 (1995)
282. K.D. Keefer, in *Silicon-Based Polymer Science. A Comprehensive Resource*, ed. by J.M. Zeigler, F.W. Gordon Fearon (American Chemical Society, Washington, DC, 1990), Chap. 13, p. 227
283. M. Beghi, P. Chiurlo, G. Cogliati, L. Costa, G. Dughi, M. Palladino, M.F. Pirini, F. Rota, in *Chemistry for Innovative Materials*, ed. by G.F. Cerofolini, R.M. Mininni, P. Schwarz (EniChem, Milano, 1991), Chap. 2, p. 31
284. D. Avnir, S. Braun, O. Lev, M. Ottolenghi, Chem. Mater. **6**, 1605 (1994)
285. D. Avnir, Acc. Chem. Res. **28**, 328 (1995)
286. J. Livage, T. Coradin, C. Roux, J. Phys.: Condens. Matter **13** R673 (2001)
287. G.F. Cerofolini, L. Meda, Appl. Phys. A **68**, 29 (1999)
288. G.F. Cerofolini, L. Meda, N. Re, in *Adsorption on Silica Surfaces*, ed. by E. Papirer (Dekker, New York, NY, 2000), Chap. 12, p. 369
289. R. Bez, E. Camerlenghi, A. Modelli, A. Visconti, Proc. IEEE **91**, 489 (2003)
290. C.D. Dimitrakopoulos, S. Purushothaman, J. Kymissis, A. Callegari, J.M. Shaw, Science **283**, 822 (1999)
291. G.F. Cerofolini, V. Casuscelli, A. Cimmino, A. Di Matteo, V. Di Palma, D. Mascolo, E. Romanelli, M.V. Volpe, E. Romano, Semicond. Sci. Technol. **22**, 1053 (2007)
292. D.C. Neckers, J. Photochem. Photobiol. A: Chem. **47**, 1 (1989)
293. C. Lambert, T. Sarna, T.G. Truscott, J. Chem. Soc. Faraday Trans. **86**, 3879 (1990)
294. A. Bandhopadhyay, A.J. Pal, J. Phys. Chem. B **107**, 2531 (2003)
295. A. Bandhopadhyay, A.J. Pal, Appl. Phys. Lett. **82**, 1215 (2003)
296. A. Bandhopadhyay, A.J. Pal, Chem. Phys. Lett. **371**, 86 (2003)
297. S.K. Majee, A. Bandhopadhyay, A.J. Pal, Chem. Phys. Lett. **399**, 284 (2004)
298. A. Bandhopadhyay, A.J. Pal, J. Phys. Chem. B **109**, 6084 (2005)
299. F.L.E. Jakobsson, X. Crispin, M. Berggren, Appl. Phys. Lett. **87**, 63503 (2005)
300. S. Karthäuser, B. Lüssem, M. Weides , M. Alba, A. Besmehn, R. Oligschlaeger, R. Waser, J. Appl. Phys. **100**, 094504 (2006)
301. R.F. Heck, J. Am. Chem. Soc. **90**, 5518 (1968)
302. K. Sonogashira, Y. Tohda, N. Hagihara, Tethraedron Lett. **44**, 67 (1975)
303. C. Frederick, Phys. Rev. D **13**, 3183 (1976)
304. B.B. Mandelbrot, *The Fractal Geometry of Nature* (Freeman, San Francisco, CA, 1982)
305. G.F. Cerofolini, Thin Solid Films **79**, 277 (1981)
306. G.F. Cerofolini, Thin Solid Films **27**, 297 (1975)
307. M. Yamaguti, M. Hato, J. Kigami, *Mathematics of Fractals* (American Mathematical Society, Providence, RH, 1997)
308. N. Rashevsky, Mathematical Biophysics, 3rd edn. (Dover, New York, 1960)
309. G.F. Cerofolini, Adv. Colloid Interface Sci. **19**, 103 (1983)
310. H.L. Ploegh, Nature **448**, 435 (2007)
311. S. Gestermann, R. Hesse, B. Windisch, F. Vögtle, in *Stimulating Concepts in Chemistry*, ed. by F. Vögtle, J.F. Stoddart, M. Shibasaki (Wiley-VCH, Weinheim, 2000), p. 187
312. Y. Gazit, D.A. Berk, M. Leunig, L.T. Baxter, R.K. Jain, Phys. Rev. Lett. **75**, 2428 (1995)
313. M.F. Shlesinger, B.J. West, Phys. Rev. Lett. **67**, 2106 (1991)
314. W.A. Steele, *The Interaction of Gases with Solid Surfaces* (Pergamon Press, Oxford, 1974)
315. D. Avnir, D. Farin, P. Pfeifer, J. Chem. Phys. **79**, 3566 (1983)
316. P. Pfeifer, D. Avnir, J. Chem. Phys. **79**, 3558 (1983)
317. D. Avnir, D. Farin, P. Pfeifer, J. Colloid Interface Sci. **103**, 112 (1985)
318. W. Friesen, R.J. Mikula, J. Colloid Interface Sci. **120**, 263 (1987)
319. K. Kaneko, M. Sato, T. Suzuki, Y. Fujiwara, K. Nishikawa, M. Jaroniec, J. Chem. Soc. Faraday Trans. **87**, 179 (1991)
320. M.W. Cole, N.S. Holter, P. Pfeifer, Phys. Rev. B **33**, 8806 (1986)
321. E. Cheng, M.W. Cole, P. Pfeifer, Phys. Rev. B **39**, 12962 (1989)
322. E.V. Albano, H.O. Martin, Phys. Rev. A **39**, 6003 (1989)
323. D. Avnir, D. Farin, P. Pfeifer, Nature **308**, 261 (1984)

324. K. Falconer, *Fractal Geometry: Mathematical Foundations and Applications*, 2nd edn. (Wiley, New York, 2003)
325. G.F. Cerofolini, D. Narducci, P. Amato, E. Romano, Nanoscale Res. Lett. **3**, 381 (2008)
326. J. Kigami, *Analysis on Fractals* (Cambridge University Press, Cambridge, 2001)
327. D. Avnir, O. Biham, D. Lidar (Hamburger), O. Malcai, in *Fractal Frontiers*, ed. by M.M. Novak, T.G. Dewey (World Scientific, Singapore, 1997), p. 199
328. H. Jan, J.M. Tour, J. Org. Chem. **68**, 5091 (2003)
329. X. Zheng, M.E. Mulcahy, D. Horinek, F. Galeotti, T.F. Magnera, J. Michl, J. Am. Chem. Soc. **126**, 4540 (2004)
330. J.D. Badji, V. Balzani, A. Credi, S. Silvi, J.F. Stoddart, Science **303** 1845 (2004)
331. Y. Shirai, A.J. Osgood, Y. Zhao, K.F. Kelly, J.M. Tour, Nano Lett. **5**, 230 (2005)
332. A. Credi, V. Balzani, S.J. Langford, J.F. Stoddart, J. Am. Chem. Soc. **119**, 2679 (1997)
333. V. Balzani, M. Gomez-Lopez, J.F. Stoddart, Acc. Chem. Res. **31**, 405 (1998)
334. V. Balzani, A. Credi, F.M. Raymo, J.F. Stoddart, Angew. Chem. Int. Ed. **39**, 3348 (2000)
335. B.L. Feringa, R.A. van Delden, M.K.J. ter Wiel, Pure Appl. Chem. **75**, 563 (2003)
336. E.R. Kay, D.A. Leigh, F. Zerbetto, Angew. Chem. Int. Ed. **46**, 72 (2007)
337. V. Balzani, A. Credi, M. Venturi, *Molecular Devices and Machines. Concepts and Perspectives for the Nanoworld*, 2nd edn. (Wiley-VCH, Weinheim, 2008)
338. W.R. Browne, B.L. Feringa, Nat. Nanotechnol. **1**, 25 (2006)
339. R. Eelkema, M.M. Pollard, J. Vicario, N. Katsonis, B.S. Ramon, C.W.M. Bastiaansen, D.J. Broer, B.L. Feringa, Nature **440**, 163 (2006)
340. M.G.L. van der Huvel, C. Dekker, Science **317**, 333 (2007)
341. J.M. Lehn, P. Ball, in *The New Chemistry*, ed. by N. Hall (Cambridge University Press, Cambridge, 2000), Chap. 12
342. N. van Gulick, New J. Chem. **17**, 619 (1993); originally presented at the *Reaction Mechanisms Conference* held in Princeton, NJ, in 1960
343. H.L. Frisch, E. Wasserman, J. Am. Chem. Soc. **83**, 3789 (1961)
344. D.M. Walba, Tetrahedron **41**, 3161 (1985)
345. J.-P. Sauvage, Acc. Chem. Res. **23**, 319 (1990)
346. R. Bez, Microel. Eng. **80**, 249 (2005)
347. R.P. Feynman, R.B. Leighton, M.L. Sands, *The Feynman Lectures on Physics* (Addison Wesley, Reading, MA, 1989), Chap. 46
348. H.B. Callen, T.A. Welton, Phys. Rev. **83**, 34 (1951)
349. G.M. Wang, E.M. Sevick, E. Mittag, D.J. Searles, D.J. Evans, Phys. Rev. Lett. **89**, 050601 (2002)
350. N. Thomas, R.A. Thornhill, J. Phys. D: Appl. Phys. **31**, 253 (1998)
351. C. Mavroidis, A. Dubey, M.L. Yarmush, Annu. Rev. Biomed. Eng. **6**, 363 (2004)
352. A. Vologodskii, Phys. Life Rev. **23**, 119 (2006)
353. R.D. Astumian, Science **276**, 917 (1997)
354. R.F. Fox, Phys. Rev. E **57**, 2177 (1998)
355. P. Reimann, Phys. Rep. **361**, 57 (2002)
356. C. Van den Broeck, P. Meurs, R. Kawai, New J. Phys. **7**, 10 (2005)
357. D. Horinek, J. Michl, Proc. Natl. Acad. Sci. USA **102**, 14175 (2005)
358. V. Serreli, C.-F. Lee, E.R. Kay, D.A. Leigh, Nature **445**, 523 (2007)
359. J.L. Lebowitz, Rev. Mod. Phys. **71**, 2 (1999)
360. J. Monod, *Le Hasard et la Nécessité* (Editions du Seuil, Paris, 1970)
361. C. Levinthal, J. Chim. Phys. **65**, 44 (1968)
362. G.F. Cerofolini, M. Cerofolini, J. Colloid Interface Sci. **78**, 65 (1980)
363. http://en.wikipedia.org/wiki/Systems-biology
364. L. Hood, J.R. Heath, M.E. Phelps, B. Lin, Science **306**, 640 (2004)
365. E. Stern, J.F. Klemic, D.A. Routenberg, P.N. Wyrembak, D.B. Turner–Evans, A.D. Hamilton, D.A. LaVan, T.M. Fahmy, M.A. Reed, Nature **445**, 519 (2007)
366. G.F. Cerofolini, G. Ferla, A. Foglio Para, Giornale di Fisica **23**, 863 (1982)

367. M. Grattarola, A. Cambiaso, S. Cenderelli, G. Parodi, M. Tedesco, B. Die, G.F. Cerofolini, L. Meda, S. Solmi, in *Molecular Electronics*, ed. by F.T. Hong (Plenum Press, New York, NY, 1989), p. 297
368. W. Aspray, IEEE Ann. History Comp. **19**(3), 4 (1997)
369. P.M. Mendes, C.L. Yeung, J.A. Preece, Nanoscale Res. Lett. **2**, 373 (2007)
370. R. Saito, G. Dresselhaus, M.S. Dresselhaus, *Physical Properties of Carbon Nanotubes* (Imperial College Press, London, 1998)
371. C. Montemagno, G.D. Bachand, Nanotechnology **10**, 225 (1999)
372. M.J. Kim, K.S. Breuer, Small **4**, 111 (2007)
373. B. Behkam, M. Sitti, Appl. Phys. Lett. **90**, 23902 (2007)
374. M. Tanaka, MRS Bull. **31**, 513 (2006)
375. R.A. Freitas Jr., J. Comput. Theor. Nanosci. **2**, 1 (2005)
376. W. Shakespeare, Hamlet, act I, scene 5, 166–7
377. P.K. Ghosh, K. Deguchi, *Mathematics of Shape Description* (Wiley, New York, 2008)

Index

About the Author

Gianfranco ("GF") Cerofolini (degree in Physics from the University of Milan, 1970) is visiting researcher at the University of Milano–Bicocca. His interests are currently addressed to the physical limits of miniaturization and the emergence of higher-level phenomena from the underlying lower-level substrate (measurement in quantum mechanics, life in biological systems, etc.).

Although his research activity has been carried out in the industry (vacuum: SAES Getters; telecommunication: Telettra; chemistry and energetics: ENI; integrated circuits: STMicroelectronics), he has had frequent collaborations with academic centers (University of Lublin, IMEC, Stanford University, City College of New York, several Italian Universities), has been lecturer in a few Universities (Pisa, Modena, and Polytechnic of Milan), and currently is lecturer at the University of Milano–Bicocca.

His research has covered several areas: adsorption, biophysics, CMOS processing (oxidation, diffusion, ion implantation, gettering), electronic and optical materials, theory of acidity, and nanoelectronics.

A gettering technique of widespread use in microelectronics, the complete setting of ST's first silicon-gate CMOS process, the development of a process for low-fluence SOI, and the identification of a strategy for molecular electronics via a conservative extension of the existing microelectronic technology are among his major industrial achievements. His main scientific results range from the preparation and characterization of ideal silicon p–n junctions and the discovery of a mechanism therein of pure generation without recombination, to the theoretical description of layer-by-layer oxidation at room temperature of silicon, and to the development of original mathematical techniques for the description of adsorption on heterogeneous or soft surfaces.

The results of his activity have been published in approximately 300 articles, chapters to books, and encyclopedic items, and in a score of patents.

GPSR Compliance
The European Union's (EU) General Product Safety Regulation (GPSR) is a set
of rules that requires consumer products to be safe and our obligations to
ensure this.

If you have any concerns about our products, you can contact us on

ProductSafety@springernature.com

In case Publisher is established outside the EU, the EU authorized
representative is:

Springer Nature Customer Service Center GmbH
Europaplatz 3
69115 Heidelberg, Germany